THE MINING DISTRICTS OF THE IDAHO BASIN AND THE BOISE RIDGE, IDAHO,

BY

WALDEMAR LINDGREN;

WITH

A REPORT ON THE FOSSIL PLANTS OF THE PAYETTE FORMATION,

BY

FRANK HALL KNOWLTON.

617

CONTENTS.

ILLUSTRATIONS.

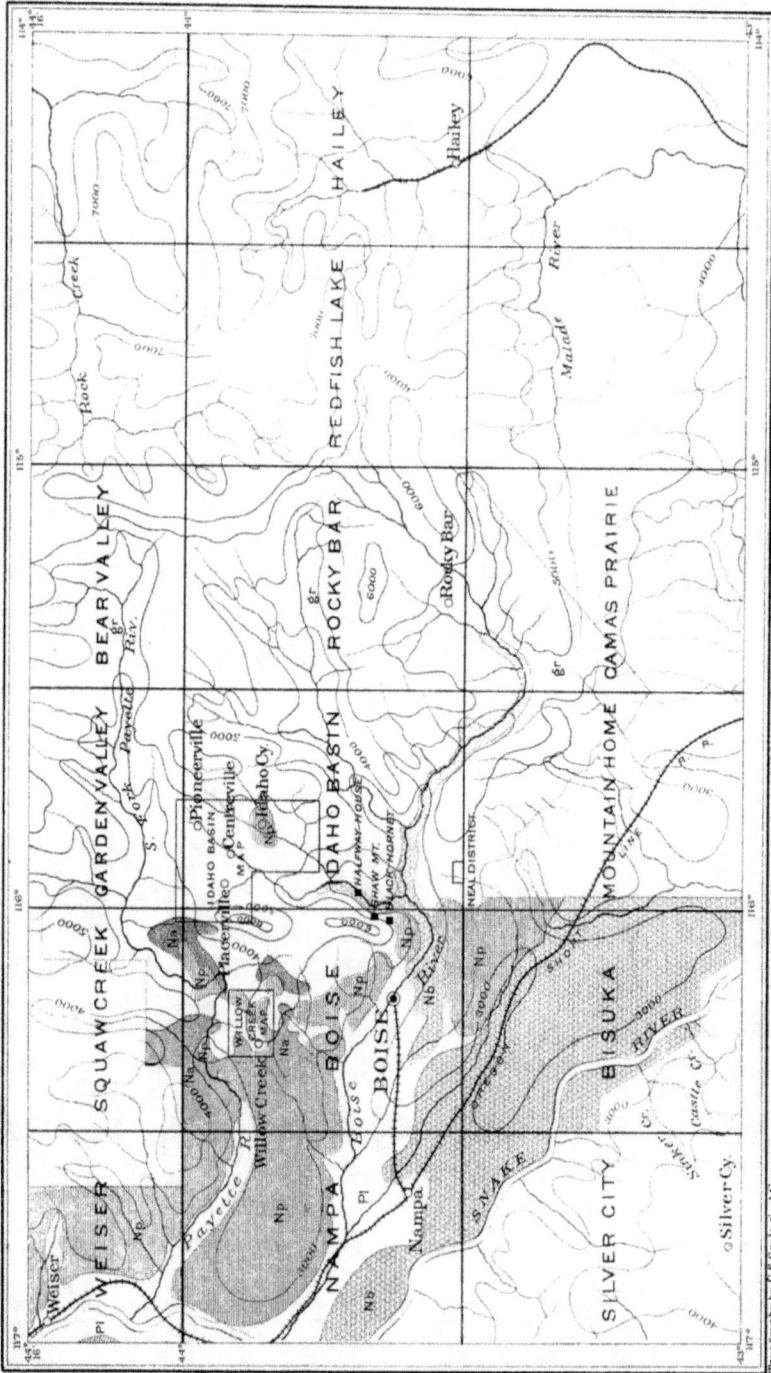

GENERAL MAP OF THE LOWER SNAKE RIVER VALLEY, IDAHO

Scale of miles

Contour Interval 1000 feet.

Topography by U.S.Geological Survey

Geology by W.Lindgren.

LEGEND

PLEISTOCENE:

Pl — SANDS & GRAVEL. River terraces.

NEOCENE:

Np — PAYETTE FORMATION Sandy Lake beds,Early Neocene

Na — BASALT AND RHYOLITE of the age of the Payette Lake beds.

Nb — BASALT (Late Neocene) with Lake beds of same age.

PRE NEOCENE:

gr — GRANITE

THE MINING DISTRICTS OF THE IDAHO BASIN AND THE BOISE RIDGE, IDAHO.[1]

By WALDEMAR LINDGREN.

CHAPTER I.

GENERAL TOPOGRAPHY AND GEOLOGY.

GEOGRAPHICAL POSITION.

The region shown in Pl. LXXXVII includes a portion of the lower Snake River Valley and the mountains adjacent on the northern side. The total area is about 13,500 square miles. It embraces, besides a part of the Snake River, almost the whole drainage of the Payette, Boise, and Wood rivers, and, in the northeastern corner, a part of that of the Salmon River. The irregular mountain complex within the drainage of the Boise and the Payette rivers is designated the "Boise Mountains;" along the parting between these rivers and the Salmon River drainage rise a series of sharp peaks, usually referred to as the "Sawtooth Range." The "Boise Ridge" extends from the Boise to beyond the Payette just west of longitude 116°, and attains elevations of over 7,000 feet. The Idaho Basin is an intermontane valley in the Boise Mountains south of latitude 44° and east of longitude 116°.

The map shows the positions and names of the quadrangles represented by the topographic sheets issued by the United States Geological Survey. It also shows, in a somewhat generalized way, the geology of a part of the area.

SNAKE RIVER VALLEY.

The discussion of the geology of this district necessitates a short reference to the large Snake River Valley and an abstract of previous work relating to its remarkable geological features. The Snake River Valley stretches across the whole width of southern Idaho in a broad curve opening toward the north and having a radius of 160 miles. The length of this valley from the base of the Tetons to Weiser, where the river enters into a narrow canyon, is over 400 miles, while its width

---

[1] The field work upon which this report is based was done during the summer and fall of 1896 by the author, assisted by Dr. E. C. E. Lord.

ranges from 50 to 125 miles, its total area being about 34,000 square miles. The grade of the river is moderate. At Blackfoot the elevation is 4,505 feet, and 350 miles lower down, at Weiser, it is 2,125 feet, giving an average grade of less than 7 feet per mile between these places. On both sides of this valley rise higher ranges, chiefly of granite in the lower valley, of granite and Paleozoic and Mesozoic rocks in the upper valley. The lower slopes of these ranges are often flanked by Tertiary lake deposits. The larger part of the valley is occupied by vast flows of basalt, frequently resting upon and covered by fluviatile and lacustrine accumulations contemporaneous with the flows. The basalt of the Snake River Valley bears evidence of being throughout of the same age approximately, though consisting of a great number of individual flows, and has generally been regarded as Pliocene. The eruptions did not originate from large volcanoes. Ashes and other fragmentary rocks are generally absent, and the basalt evidently flowed out quietly and without explosions from numerous local vents along the margin of the valley or within the valley itself. This volcanic action is usually referred to as fissure eruption, but it must not be inferred that these fissures were long or large. It appears rather as if the vents had the character of rounded local orifices, hardly extensive enough to be classed as fissures. The basalts often flowed down from the foothills of older rocks, closely following the present canyons, though the streams have since then generally succeeded in wearing through the filling in their bottoms. Thus it is, for instance, along the Boise River.

It will be shown that the Snake River Tertiaries consist of a thick series of early Neocene (Miocene) lake beds, with which are associated vast masses of eruptives distinct from the Snake River basalts proper, and another series of deposits of late Neocene age (Pliocene), consisting of the Snake River basalts and associated sedimentary rocks. These two terranes represent successive stages of the Neocene lake and are often difficult to separate.

LITERATURE.

The upper Snake River Basin has been described by Messrs. Hayden,[1] Bradley,[2] Peale,[3] and St. John,[4] in the reports of the United States Geological and Geographical Survey of the Territories. Hayden describes the basalt flow along the present line of the Utah and Northern Railroad, mentioning that there were several flows of basalt, or at least two, separated by somewhat tilted Pliocene deposits of slight depth. St. John and Peale describe the basalt flows east of this, near the headwaters of the Snake River. Peale states that a number of extinct craters exist, that there were several flows of

---

[1] U. S. Geol. and Geog. Surv. Terr., Rept. 1871, pp. 25-30.
[2] Ibid., Rept. 1872, p. 190.
[3] Ibid., Rept. 1877, p. 548.
[4] Ibid., Rept. 1877, p. 323.

basalt, and that the basalts are generally horizontal in position and fill the valleys and the more depressed portions of the basins. There appear to have been two periods of basaltic flows, one at the close of the Pliocene, the other at the beginning of the Pleistocene. The Pleistocene age is inferred from exposures at Marsh Valley, near Red Rock Pass, where Pleistocene beds were somewhat eroded before the basaltic flow.

According to Gilbert, however, this Pleistocene is older than the highest stage of Lake Bonneville, during which the lake found an outlet at Red Rock Pass. The river draining the lake at this time appears to have flowed over the surface of the basalt.

According to Hague,[1] the latest eruptions in the Yellowstone National Park are of basalts, which stretch far into Idaho in somber, monotonous beds. These basalts are pre-Glacial, and their eruption is referred to the Pliocene.

About 1869 Mr. Clarence King visited the lower part of the Snake River basin and collected a number of fossils from beds beneath the basalt at Castle and Sinker creeks, tributaries from the south, joining the river about due south of Boise. The fossils have been described in detail, while no description of the localities was ever published, a fact which has led to some confusion. A few notes regarding this occurrence are contained in King's Systematic Geology[2] and may be quoted:

In the basin of Snake River . . . there were basaltic eruptions in the middle of the Pliocene period which overflowed the earlier lacustrine beds of the period, and in turn were themselves overlaid . . . by the main, later Pliocene series. . . . Sections obtained along the plains between the Owyhee Mountains and Snake River show that a considerable portion of the beds of the valley, which consist chiefly of white sands and marls carrying numerous well-defined Pliocene forms, were overlaid by large accumulations of basaltic flow, and that subsequently a second period of lacustrine deposition took place, likewise characterized by Pliocene forms, the latter representing a more advanced stage of development and more recent type than those beneath the basalt.

King further states that near Shoshone Falls the basalt rests on the eroded surface of a trachytic soft rock which he considers of pre-Miocene age.[3] From the collections of King and the later collections of Wortman, Cope has described an extensive fauna of fresh-water fishes, and proposed for the sediments in which these are contained the name *Idaho formation*.[4]

The locations given are very vague, as "Catherine Creek," "Castle Creek," or "Southern Idaho," and no description of the beds is vouchsafed. The fauna consists of 22 species, all differing from existing species so far as known. Professor Cope thinks that the

[1] Am. Jour. Sci., 4th series, June, 1896, Vol. I, p. 455.
[2] U. S. Geol. Expl. Fortieth Par., Vol. I, 1878, pp. 418, 440.
[3] Ibid., p. 593.
[4] Proc. Phila. Acad. Sci., 1883, pp. 153–166.

evidence clearly indicates a Pliocene age. From the same beds were obtained three species of crawfishes specifically distinct from all others described from the West. Mammalian remains were also collected by King from similar beds on Sinker Creek, which were determined by Leidy to be *Mastodon mirificus* and *Equus excelsus*, both of which belong to the Niobrara Pliocene fauna. A few mollusks have also been found in the same deposits on Sinker Creek. Thus Meek[1] described *Sphaerium* (?) *idahoense* Meek from Castle Creek, collected by King. Gabb[2] described two species, *Melania taylori* Gabb and *Lithasia antiqua* from a "Deposit on Snake River on the road from Boise to the Owyhee mining country;" thus probably from Walters Ferry. The same forms have been found, according to Mr. George H. Eldridge, at Glenns Ferry, 120 miles above Walkers Ferry. Dr. White describes the same two species and another, *Latia dallii*, from a point 50 miles below Salmon Falls, Snake River, which probably refers to Glenns Ferry, and states that these forms differ considerably from any known fresh-water fauna of America either fossil or living.[3] Both Meek and White are in favor of correlating these Tertiary beds with King's Truckee Miocene. To this the utterances of King are directly opposed, and it is, indeed, from stratigraphic grounds, improbable that these beds are of Miocene age. Near Glenns Ferry beds of sand and clay occur intercalated between the basalt flows, and it is probable that the fossils came from this locality and that all of them were collected in beds very closely associated with the late basaltic eruptions, from which it would follow that they should be placed in the latest Pliocene.

Prof. O. C. Marsh states (oral communication, January, 1897) that a large amount of Pliocene mammalian remains was found in a bluff about 100 feet above the Snake River, some distance below Weiser, at the old crossing of the stage road to Oregon, on the Oregon side of the river.

None of these localities were visited during the field season of 1896 on account of pressing economical and areal work in other sections; but from the area studied it was possible to read in its chief features the later geologic history of the lower part of the Snake River basin. The correlation of these results with the older work remains for the future.

TOPOGRAPHY.

The chief topographic features of the region, of which the geology is shown on Pl. LXXXVII, are as follows:

Broad flat mesas of basalt and Pliocene lake beds spread on both sides of Snake River, though chiefly on the northern side. Through these mesas the river has cut an abrupt canyon, bordered by basaltic

---

[1] Proc. Phila. Acad. Nat. Sci. 1870, p. 57.
[2] Paleontology of Cal., Vol. II, p. 13.
[3] Proc. U. S. Nat. Mus. 1882, Vol. V, p. 99.

cliffs, to a depth of from 200 to 700 feet. The low mesas, surmounted by several buckles or domes of basalt a few hundred feet high, rise gradually toward the edge of the mountain. Near the mouth of the Boise River the basalt mesas cease, and from here down to Weiser, where the great Snake River canyon begins, several large tributaries enter, such as the Payette and the Owyhee, and, at elevations of from 2,200 to 2,700 feet, level bottom lands and broad low terraces flank the water courses.

Between the mouth of the Boise and Weiser flat-topped hills of soft sandstones rise on both sides of the Snake River to a height of 600 to 800 feet. Similar complexes of high sandy mesas rise between the lower courses of the Boise and the Payette and north of the Payette. The mountains of older rocks surrounding the tectonic trough of the Snake River Valley rise gradually, on the north side of the river, beyond the sloping mesas of Tertiary rocks, their margin having a northwesterly direction in this vicinity. The transition between mountains and mesa is abrupt only at the Boise River, near Boise, and the abruptness is here due to the extensive erosion of the Payette sandstones by the river.

The mountain region extending up to the Sawtooth Range, dividing with a north-northwesterly trend the waters of the Boise and the southern branches of the Payette from those of the Salmon, has an average width of 55 miles and culminates in summits with an elevation of from 10,000 to 11,000 feet. This mountain complex, which is made up chiefly of granitic rocks, does not form a well-defined range, but rather a broad uplift dissected deeply and in the most intricate manner by the forks of the Boise and the Payette. The summits of the narrow ridges generally form gently sloping lines. If a surface were constructed containing all these lines it would be of undulating, curved character, sloping gently from elevations of 9,000 down to 4,000 feet. From the southwestern edge a steeper slope carries the granitic rocks below the surface of the Tertiary rocks of the Snake River Valley. The canyons of the Boise and the Payette have cut down in the uplift to a maximum depth of 3,000 feet, and are joined by deep lateral canyons, dividing the whole region into a maze of narrow arêtes. The grade of the main rivers is relatively low, from 10 feet up to 50 feet per mile, and only well up toward the head waters are grades of 100 feet per mile attained. The grades of the lateral canyons are also often relatively slight in their lower course, but extremely steep cirques rise near their head waters. The Idaho Basin quadrangle offers excellent illustrations of these relations, which are the result partly of a considerable antiquity of the drainage, partly of the crumbling character of the granite. At the main divide (Bear Valley quadrangle) the broad valleys and gentler slopes of the Salmon River drainage contrast strongly with the deeply incised canyons of the Boise and Payette. The latter streams are continually capturing territory belonging to the former, and the divide is in process of migration north-eastward. The whole region may be regarded as an uplifted sloping plateau deeply dissected by a drainage system, whose origin evidently antedates the Miocene period. Smaller individual ranges occur in a few places, as in the Boise Ridge, rising to elevations of 7,500 feet and extending due north, dividing the Idaho Basin from the waters of the Payette. This range also crosses the South Fork of the Payette and continues for some distance north of it. Within this mass of mountains several depressions or basins with gentler slopes also exist, such as the Idaho Basin, the Deadwood Basin, and Smiths Prairie, which have been created or emphasized by more recent orographic movements. Evidences of glacial topography occur only near the Sawtooth and Trinity mountains.[1] The lower area here specially described has never been covered by ice.

GEOLOGICAL HISTORY.

The vicinity of Boise River, where it debouches from the mountains, proved to be an exceptionally fortunate location for the study of the geological history of this part of the Snake River drainage, for the record left by the river of successive geological events back to a certain date is remarkably clear and easy to read.

PRE-TERTIARY.

The oldest rock exposed is the granite of the Boise Mountains. This forms an extremely large area, embracing, so far as known, the whole of the upper drainage of the Boise and Payette rivers and extending northeastward beyond the Sawtooth Mountains and eastward as far as Wood River, where it is adjoined by sedimentary rocks of probably Carboniferous age.[2] This rock is largely a typical, coarse granite of gray or yellowish-gray color, consisting of orthoclase in often large crystals, plagioclase, quartz, biotite, and sometimes muscovite. Pegmatite dikes are common in many places. Locally the granite contains hornblende, and is by gradual transition connected with intermediate rocks standing between granite and diorite, and also, though more rarely, with diorites. Narrow dikes of light-colored granite-porphyry and dark lamprophyric dike rocks, chiefly minettes, are abundant and present a great variety of structural types. A belt characterized by dikes of coarse quartz-diorite, porphyrites, and occasional occurrences of gabbro and diabase extends, with one short interruption, from the vicinity of Wilson Peak, east of the Idaho Basin, by Quartzburg, to the Willow Creek mining district. All of these dikes are probably connected with the granite eruption—that is, they were intruded shortly after the consolidation of the granite. Within the area described the granite is remarkably unaltered and massive, no

---

[1] George H. Eldridge, Sixteenth Ann. Rept. U. S. Geol. Survey, Part II, 1895, p. 223.
[2] George H. Eldridge, loc. cit.

bodies of schist appearing in connection with it. It weathers easily and crumbles to a coarse sand which largely covers the hillsides. Only in the higher mountains and along the bottom of the canyons are satisfactory exposures found. The age of this granite, which is clearly of igneous and intrusive origin, is an open question. Messrs. Becker[1] and Eldridge[2] assign to it provisionally an Archean age, but a thorough study of its contact with surrounding formations is necessary before its age can be determined. The granite is in many places traversed by shear planes, giving it a jointed or sheeted structure, and often these planes coincide with the direction of the fissures on which mineral veins occur. It is probable that these two features are of the same and contemporaneous origin. Nearly all of the primary mineral deposits are contained in the granite or allied porphyries. By far most of them have a direction ranging from E.-W. to ENE.-WSW., and dip to the south at angles from 40° to 85° from the horizontal. While it is probable that all of them belong to the same period of formation, there are few definite clews to their age. It is likely, however, that they are post-Carboniferous, and it is certain that they antedate the Miocene lake deposits. A Cretaceous or early Tertiary age may provisionally be assigned to them. The mode of their occurrence indicates beyond doubt an origin by deposition from mineral waters, probably ascending hot springs. A slight recurrence of the vein-forming activity occurred after the Neocene period.

Before the beginning of the Neocene the chief features of the topography were outlined—the broad uplift of the Boise Mountains and the depression of Snake River Valley. The latter is not unlikely a sunken area separated by old fault lines from the mountains to the north. At that time the basalt flows and the lake beds did not exist, but the drainage of the Boise, and probably also of the upper Payette River, was outlined in practically its present form. The granitic range presented a bold scarp facing the valley, and the canyon of the Boise River was, at its debouchure from the mountains, cut to practically the same depth which it has at present. It had not, of course, cut back so far toward the Sawtooth Range as at present, and many features of the drainage, notably in the Idaho Basin, were different from those existing now. As substantiating this it will be shown that the Miocene lake beds fill the old canyon at the gate of the mountains, 10 miles southeast of Boise, and that in front of it lie enormous masses of coarse Neocene gravel and conglomerate. Thus the time immediately preceding that from which the first records date was one, first, of uplift and subsidence, during which the rough features were blocked out, and second, one of long-continued erosion, during which the Boise Mountains were dissected and the débris from the excavated canyons deposited in the basin of the Snake River Valley, where it is

---

[1] Tenth Census, Precious Metals, p. 54.
[2] Loc. cit.

now deeply covered below later formations. If we should venture tentatively to go back one step further, it might be suggested that the uplifted surface of the Boise Mountains is probably the result of a far older erosion, of early Tertiary or Cretaceous age, which planed down a more ancient range to gentler outlines, or to a peneplain.

THE PAYETTE FORMATION.

During the earlier part of the Neocene (Miocene) a large fresh-water lake occupied at least the lower part of the Snake River Valley, and its sediments are now prominent features of the region. For these lake beds the name Payette formation is proposed, and their age is determined as upper Miocene. This formation is probably not the same as Cope's Idaho formation, to which a Pliocene age was assigned and which appears to be connected with the later basalt flows.[1]

The extent of the formation is shown in Pl. LXXXVII, from which it is seen that it lies in front of the Boise Mountains and occupies the whole lower part of the ridge between the Boise and the Payette. It extends over large areas to the north of the Payette, along the flood plains of the Snake River, and is seen to occupy vast areas in Oregon between the mouth of the Owyhee River and Weiser, where the Snake River Canyon begins. On both sides of the lower Snake River the bluffs of the Payette formation attain a height of over 800 feet. In the Payette Valley south of Emmett the sharply defined bluff of Payette beds rises 600 feet above the alluvium. Smaller masses, detached by erosion or uplifts, lie in the intermontane valleys as far east as the Idaho Basin.

Along the Boise Mountains the Payette beds rest against the irregularly eroded and sharply sloping surface of the granite, and the top stratum attains a height of 4,100 feet. A total thickness of 1,000 feet is exposed near Boise, and wells bored show several hundred feet of similar strata below the surface. Over the larger part of its extent the formation lies nearly horizontal or dips only a few degrees. Near the mountains dips of 8° to 10°, generally westward, are noted, and the smaller detached masses in the intermontane valleys are still more disturbed, generally dipping westward at angles up to 50°. This is particularly marked in the long arm of sediments of the Payette formation filling the valleys of Horseshoe Bend and Jerusalem, on the Payette.

As might be expected from the character of the land mass from which the sediments were obtained, the latter consist chiefly of granitic, light-colored sands, locally cemented by hot-spring deposits to hard

---

[1] From the results of more extended surveys during the summer of 1897, it has become evident that the Payette and Idaho formations represent two successive stages of the lake, the often deformed shore-line of the former being found at elevations of from 4,200 to 5,000 feet, and those of the latter at a maximum elevation of 3,000 feet. To separate the deposits of the two formations is not always easy.

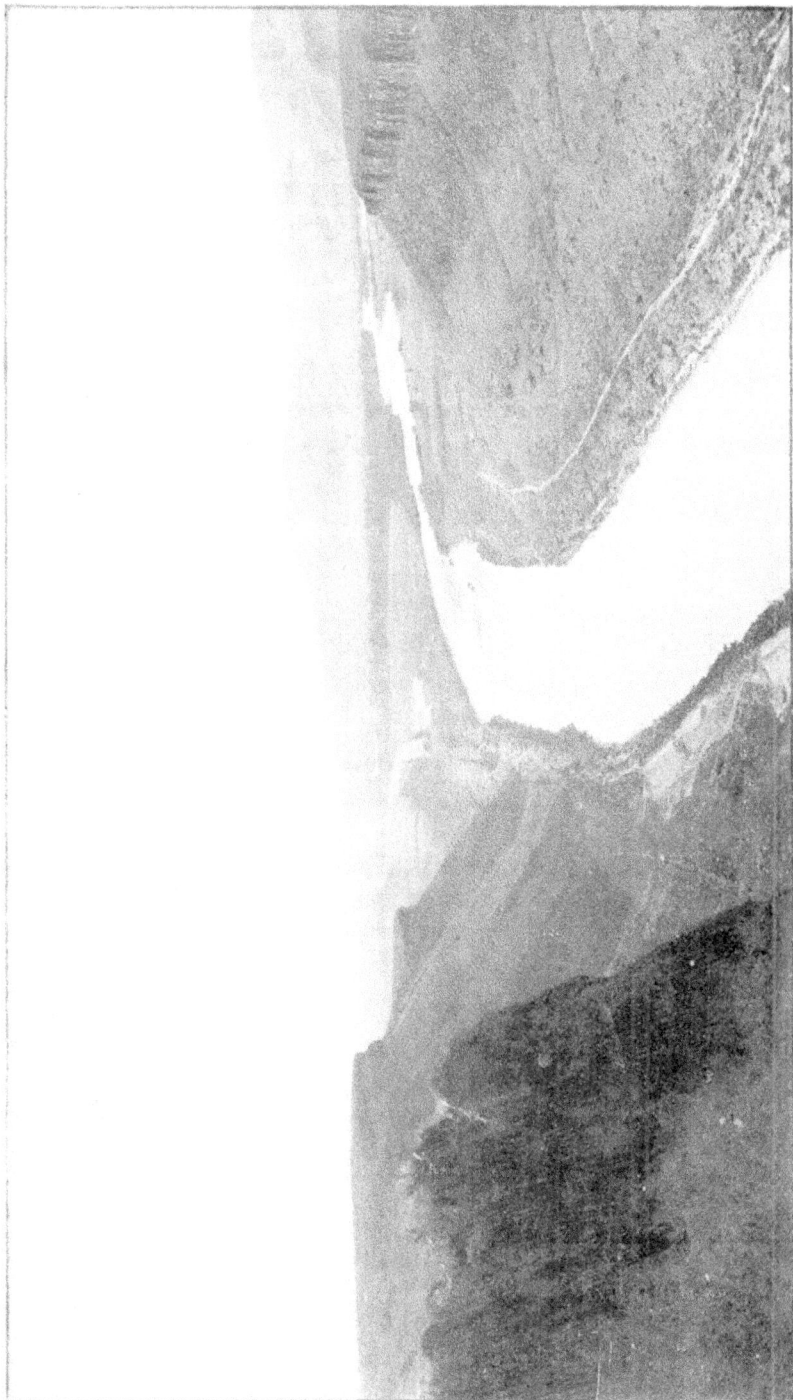

HEAD OF BOISE VALLEY, 8 MILES SOUTHEAST OF BOISE, LOOKING NORTHWEST, SHOWING BASALT TABLE AND PLEISTOCENE TERRACES.

sandstones (as at Table Mountain near Boise; see Pl. LXXXVIII) or clayey semiconsolidated sandstones. Heavy masses of conglomerates and gravels begin to appear at Table Mountain, and reach their greatest development opposite the mouth of Boise River, in the high ridge extending in a westerly direction. Purely clayey deposits are rarer, occurring only in convenient sheltered locations near the shore line or in places where volcanic eruptions took place. The basal part of the formation contains, at Horseshoe Bend, Jerusalem, and other localities along the Payette, small coal seams. In the clay accompanying these coal seams vegetable remains are of frequent occurrence. The following forms were identified by Professor Knowlton:

Two miles southeast of Marsh post-office, on the Payette, is a small coal prospect. The disturbed beds of the Payette formation lie here on a sharply sloping surface of granite. At this place the following forms were found, together with many new species:

> Sequoia angustifolia ? Lx.
> Quercus consimilis Newb.
> Acer trilobatum productum ? Heer.

One mile southeast of Marsh, along the road to Willow Creek, a few hundred feet above the river, is an outcrop of yellowish-gray "chalk," or diatomaceous earth, intercalated in the Payette beds and capped by basalt. In this the following forms were found, in an excellent state of preservation:

> Salix angusta Al. Br.
> Quercus consimilis Newb.
> Q. simplex Newby.
> Platanus aspira ? Newb.

From Cartwright's ranch, on Shafer Creek, 4 miles southeast of Horseshoe Bend post-office, the following forms were identified, together with several new species. This is an excellent locality, and the leaves embedded in a dark clay shale are beautifully preserved.

> Sequoia angustifolia Lx.
> Salix angusta Al. Br.
> Ficus ungeri Lx.

This area is separated from the main one by a small ridge of granite, but that the two formations are identical admits of no doubt. The beds are here disturbed, dipping west at 20° to 25°. The plants were obtained near the base of the series, of which about 800 feet of alternating, fairly consolidated shale and sandstone are exposed. The elevation is 3,500 feet.

Near Idaho City another detached mass of lake beds is preserved at an elevation of from 4,000 to 4,500 feet, the occurrence of which is described more in detail in Chapter III. From the middle of a series 800 feet thick the following plants were obtained:

> Sequoia angustifolia Lx.
> Ulmus speciosa Newberry.
> Betula angustifolia Newberry.

From these data Professor Knowlton draws the conclusion that the age is Upper Miocene, contemporaneous with the flora of the auriferous gravels and the Ione formation of California, the Lamar flora of the Yellowstone National Park, and the John Day formation of Oregon. The paleobotanical evidence confirms the conclusion, confidently drawn from the field work, that all these smaller detached masses of lake beds are of practically the same age.

During the time of the maximum extension of the Payette Lake its surface stood at the present elevation of 4,200 feet. Its deposits, over 1,000 feet thick near the shore, rested against the abrupt slope of the Boise Mountains, and filled the old canyon of the Boise to the same depth. The canyon must have formed a fiord, the branches of which reached as far back as the Idaho Basin, and in which vast quantities of gravel and sand accumulated. Isolated occurrences of well-washed gravel on the summit of high ridges in the lower Moore Creek drainage, at elevations of 4,500 feet, confirm the above conclusions. The data are not at present sufficient to determine the extent of the Payette Lake, though it is probable that it was confined to the Snake River Valley, inclosed on the west by the Blue Mountains and on the east by the divide toward the Salmon River.

EARLY NEOCENE VOLCANIC ACTIVITY.

Near the base of the Payette formation sheets of rhyolite and rhyolite-tuff occur, but this eruption was of limited extent. The best exposures are found near Boise and in the Willow Creek mining district. After the rhyolitic eruptions there occurred enormous outpourings of basaltic lavas, distinctly different from and earlier than the Snake River basalts. The rocks are in part normal basalts, but have usually a somewhat andesitic habit. In large areas the outcrops generally have a reddish-brown color, distinct from the somber black of the later flows. These early Neocene eruptives are to some extent represented near Boise, but become more abundant northward. Many exposures are seen along the lower Payette from Marsh post-office to Horseshoe Bend and Jerusalem Valley, and the conspicuous sharp ridge of Squaw Butte, rising just north of the Payette above the lake beds to an elevation of 5,800 feet, is composed entirely of these older effusive rocks. Squaw Butte is well visible from the railroad near Nampa and Caldwell as a rugged, reddish-brown peak, contrasting with the white lake beds. The evidence shows that these flows were contemporaneous with the deposition of the Payette beds, and are underlain and covered by sandstones. In some places they break through the lower lake beds and metamorphose them. The vents were located chiefly along the margin of the old lake. At Jerusalem (see Pl. LXXXIX) and in Squaw Butte the whole volcanic series—in the latter place consisting of countless flows, attaining a thickness of several thousand feet—has been disturbed and uplifted, and now dips to the west at angles ranging from 10° to 35°, the monoclinal uplifts vividly recalling those in eastern Washington described by Professor Russell.[1] There, however, the Miocene beds lie only on top of the volcanic flows of the Columbia formation, while here they lie both below and above similar igneous masses.

On both sides of the Snake River, from Caldwell to Weiser (near the upper edge of the map, Pl. LXXXVII), the beds do not contain much evidence of volcanic activity, though from Parma a few volcanic buttes are visible from the railroad, far to the west, among the lake beds in Oregon. But a few miles beyond Weiser the white bluffs of lake beds begin to assume red or orange colors, and contain streaks of intercalated tuffs. The valley contracts, and at the entrance of the canyon basaltic rocks appear, partly interstratified with the lake beds, partly underlying them. Near Huntington the deep and narrow canyon is composed entirely of basic volcanic rocks, clearly connected with the early Neocene lake beds, and assuredly different from the later basalts from the upper valley. How far down this volcanic canyon extends is not known.

POST-PAYETTE EROSION.

After attaining its highest stage, the lake was drained by the establishing of the present course of the Snake River below Weiser. The lake receded as the canyon was rapidly cut by the mighty volume of water, and erosion has steadily proceeded since the end of the Miocene or the beginning of the Pliocene. The broad valleys of the Boise, the Payette, and the lower Snake were eroded in the soft lake beds. A new course was established for the Payette River, which evidently did not debouch at its present position before the Payette epoch. The Boise River, on the contrary, maintained its old course. The accumulated gravels were scoured out from its canyon, and, before the Pliocene basaltic eruptions, its channel in the canyon was deepened nearly to its present level. There was, however, at least one temporary check in this process of draining. For a considerable interval of time the lake remained stationary, at a present elevation of from 2,800 to 3,000 feet. The deposits and basalt flows of this epoch are regarded as late Neocene (Pliocene) and belong to the Idaho formation of Cope.

POST-PAYETTE OROGRAPHIC DISTURBANCES.

Before the epoch of the Pliocene basalt flows the sediments and flows of the Payette formation were subject to some disturbances, reaching their maximum in the smaller areas in the intermontane valleys. Certain parts of the series acquired a slight westerly dip. More intense orographic movements took place at Squaw Butte and in the Horseshoe Bend and Jerusalem valleys, resulting in monoclinal

---

[1] Bull. U. S. Geol. Survey No. 108, 1893, p. 28.

uplifts, the detailed character of which will be discussed in the text accompanying the Boise folio of the Geologic Atlas of the United States. Important movements also took place along the Boise Ridge, which separates the Idaho Basin from the Payette River, and it is probable that it has undergone an extensive uplift. Over the larger part of the area no orographic movements have affected the beds.

LATE NEOCENE BASALTS.

When the lake had been partly drained, vast basaltic eruptions began, and in time, intercalated with lake beds, filled the whole of the upper Snake River Valley from the base of the Tetons, near the Wyoming line, to a point near the confluence of the Boise and the Snake. Between this point and Weiser no basalts are seen. The basalt flows lie horizontal, filling the plains and the modern canyons. They are also distinguished by their fresh character, black color, and columnar structure. The aggregate thickness probably never exceeds 1,000 feet, and is ordinarily much less, individual flows being rarely over 100 feet thick.

As already indicated, the basalts were erupted from a great number of inconspicuous craters, both in the plains and in the adjoining mountains. Their fluidity was remarkable, continuous flows of 50 miles or more being noted. Where the Boise River emerges from the mountains the exposures are exceptionally good. There are three or four flows, the principal ones coming down from the South Fork of the Boise. The deepest flow of comparatively small volume is probably the oldest, and lies but a few feet above the present bed of the river. Deep river gravels accumulated on this flow, and, soon afterward, two later flows came down the Boise River and filled the canyon near the mouth to a depth of 300 feet. Beyond the mouth the basalt spreads out, and its surface rapidly sinks westward. Still another basalt flow, about 75 feet thick, came down from Moore Creek and joined the large ones at the mouth of the main river. Above the source of this basalt the damming resulted in terraces and bench gravels now lining the upper valleys of Moore Creek and Grimes Creek, described in Chapter III.

POST-BASALTIC EROSION.

If the epoch of the basaltic flows be placed at the very close of the Neocene, the events that have taken place since then must be referred to the Pleistocene. Among these are the erosion of the canyons of Snake River and its tributaries to a depth of from 200 to 700 feet and the deposition of extensive flood plains and terraces along the lower Snake, Boise, and Payette. The Boise River has, in Pleistocene times, cut through the 300 feet of basalt accumulated at the mouth of its canyon, and thus laboriously regained the same stage it occupied before the beginning of the Payette epoch. The direction of the channel has gradually changed. During the Payette epoch it had a nearly

UPPER CANYON OF BRAINARD CREEK, JERUSALEM VALLEY, BOISE RIDGE, SHOWING TILTED BASALT FLOWS RESTING ON GRANITE.

westerly trend, while subsequent events have more and more forced it in a northwesterly direction. Pl. LXXXVIII illustrates well the present condition at the mouth of the canyon—the basalt flow cut in two and the two Pleistocene flood plains in the widening valley of the river. Pliocene gravel older than the flood plains underlies the basalt. In the right background the Miocene sandstones of Table Mountain are shown, while the background at the extreme right shows the first granitic hills of the Boise Ridge.

Having thus traced the cycle of events which have taken place in this region since the beginning of the Neocene, it may be well to rapidly review the main points in the history.

RÉSUMÉ OF GEOLOGICAL EVENTS IN THE LOWER SNAKE RIVER BASIN.

Pre-Neocene	{ The depression of the Snake River Valley outlined by orographic movements.
Pre-Neocene or early Neocene (Miocene).	{ Long-continued erosion, dissecting the Boise Mountains. Boise Canyon eroded to its present depth at its mouth.
Early Neocene (Miocene), possibly extending over into late Neocene.	{ Large fresh-water lake occupying Snake River Valley. Deposition of the Payette formation at least 1,200 feet thick. Highest level reached, 4,200 feet above present sea level. Eruptions of rhyolite and andesitic basalts, contemporaneous with the sedimentation.
Late Neocene (Pliocene)	{ Orographic disturbances of the Payette formation. Partial drainage of the lake and epoch of erosion excavating the valleys of the Snake, Boise, and Payette from the Payette lake-beds. Basalt flows, filling the Snake River plains and the Boise Canyon. Deposition of lacustrine sediments between the flows. (Idaho formation.) Complete draining of lake.
Pleistocene	{ Erosion of the Snake River basalt canyon above the confluence with the Boise; Boise River cuts down through the basalt to its present depth. Terraces, up to 100 feet above the river, and present flood plains along the Lower Snake, Boise, and Payette rivers.

CHAPTER II.

THE ORE DEPOSITS IN GENERAL.

GENERAL FEATURES.

Throughout the Boise Ridge and the Idaho Basin the primary gold deposits present a certain similarity. They are all contained in granitic rocks or associated dikes. They are all either fissure veins or impregnations connected with fissures. Nearly all of these fissures have a direction ranging from east-west to northeast-southwest, the chief exceptions to this rule being found in the Black Hornet district. The dip is ordinarily to the south at angles of from 45° to 89°, but in the Willow Creek and Rock Creek districts similar dips to the north are found. The prevailing direction of the fissures is the same as that of an often well-developed system of joint planes commonly seen in the Boise Ridge. Finally, the ores are, on the whole, of a similar character, consisting chiefly of auriferous pyrite, arsenopyrite, zincblende, and galena in a gangue of quartz or, more rarely, calcite. The fresh ores from deeper levels contain a variable percentage of free gold. Rarely, however, is more than 60 per cent of the total value caught on the amalgamating plates. Gold predominates largely in the value of the ore, though seldom by weight, for in the ordinary ores the weight of the silver considerably exceeds that of the gold. The alteration of the country rock in the vicinity of the veins is throughout of the same character.

ALTERATION OF THE COUNTRY ROCK.

A marked change appears in the rock in the vicinity of the veins. The dark constituents, biotite and hornblende, are bleached or disappear, and the feldspar is altered to a soft, white, opaque material, only the quartz remaining unaltered. Besides abundant iron pyrite, arsenopyrite also appears in small, scattered, perfect crystals. The width of the altered zone may be from a foot up to 50 or 60 feet. This alteration of the country rock has been noted by Messrs. George H. Eldridge,[1] J. B. Hastings,[2] and F. D. Howe, but has ordinarily been described as kaolinization. The soft, white substance, which often has a greasy feel, is also referred to by the miners as "talc." This change in appearance and composition is without the slightest

---

[1] Sixteenth Ann. Rep. U. S. Geol. Survey, Part II, 1895, pp. 225, 252.
[2] Idaho Mining News, Vol. L, No. 1, p. 15.

doubt directly due to the chemical action of the solutions from which the mineral content of the vein was deposited. The process consists in a sericitization, or replacement of the ferromagnesian silicates, feldspar, partly also the quartz, by sericite, a fine-fibrous or felted variety of white mica. In composition it is a hydrous silicate of aluminium and potassium, probably identical with muscovite. In many places a carbonatization, or replacement by carbonate of lime and magnesia, goes on at the same time, and sulphides, chiefly pyrite and arsenopyrite, rarely other minerals, are usually also formed in the rock as minute and perfect crystals. It is certain that this metasomatic process is a common one in fissure veins,[1] and its chemical character is very different from kaolinization. Kaolin, in fact, is a product not ordinarily found on the mineral veins, and talc occurs still more rarely.

This altered granite, together with the ore and gangue occurring in seams or veins through it, constitutes what miners term a "ledge" and "ledge matter." This may be many feet wide, and the paying ore may, and in fact does usually, form only a small portion in width of the ledge matter. The altered country rock, though often well filled with pyrite and arsenopyrite, is ordinarily very poor, containing at most one or two dollars in gold. Exceptionally it contains enough gold to be considered an ore, but generally only when adjoining rich vein filling and large ore shoots; even then it constitutes only second-class ore, and its sulphides are much poorer than those in the vein proper. It rarely contains any free gold. As examples a few typical rocks of this kind may be described.

The altered dioritic granite from the Checkmate mine, Willow Creek (88 Boise sheet collection), is a granular, white, soft rock, consisting of quartz grains, white earthy grains replacing the feldspar, a few foils of pale-yellowish mica and abundant small and perfect crystals of pyrite, showing the combination $\infty 0 \infty$, $\infty 0 2$, and a few small prisms of arsenopyrite. A few narrow seams, 1^{mm} wide, traverse the rock and carry only blende and galena. The microscope shows quartz grains with undulous extinction, which contain in places a few shreds of sericite, but are on the whole hardly attacked by any metasomatic process. There are a few larger muscovite foils, which evidently represent the biotite of the fresh rock. The space between the quartz grains is filled by a fine-felted sericite mass, which incloses nearly all of the idiomorphic pyrite. Intergrown with and inclosed by the pyrite is a little brown zinc blende and galena in small anhedral grains; one crystal of arsenopyrite was also noted. No calcite was found. This rock occurs close to a rich ore body, but an assay of it gave only 0.1 ounce of gold and 0.5 ounce of silver per ton. The total replacement

---

[1] For studies of the metasomatic processes of fissure veins by the author, see Fourteenth Ann. Rept. U. S. Geol. Survey, Part II, pp. 243-284, and Seventeenth Ann. Rept. U. S. Geol. Survey, Part II, p. 144; also Bull. Geol. Soc. Am., Vol. VI, p. 221.

of the feldspars and ferromagnesian silicates and the immunity of the quartz is noteworthy.

From the Silver Wreath tunnel, Willow Creek district, Boise County, two rocks were collected, analyses of which are given below. The first is a perfectly fresh granitic rock, the ordinary country rock of the district. The second, occurring only a few feet away, is the same granite altered by the vein-forming agencies, and a comparison between the two will show the character of the process. Both rocks are unaffected by atmospheric agencies.

Analyses of rocks from Silver Wreath tunnel.

[Analyst, George Steiger.]

	I.	II.
SiO_2	65.23	66.66
TiO_2	.66	.49
Al_2O_3	16.94	14.26
Fe_2O_3	1.60	.67
FeO	1.91	2.41
MnO	trace	trace
CaO	3.85	3.37
BaO	.19	none
MgO	1.31	.95
K_2O	3.02	4.19
Na_2O	3.57	none
H_2O below 100°	.18	.36
H_2O above 100°	.88	2.16
P_2O_5	.19	.17
SO_3	none	none
S	none	.95
CO_2	.25	3.67
	99.78	100.31
Less O		.24
		100.07

I. 79 Boise collection; fresh granitic rock.
II. 80 Boise collection; altered granitic rock.

The unaltered rock is light gray and coarse grained, the average size of the constituents being 5^{mm} to 6^{mm}. With the naked eye, white plagioclase, reddish orthoclase, biotite, titanite, and quartz may be distinguished. Under the microscope the quartz proves to be very abundant and is slightly crushed. The plagioclastic feldspars predominate and occur generally in anhedrons, more rarely in roughly prismatic forms. Symmetric extinctions show a maximum of 10°. A little orthoclase lies between the plagioclase grains. The biotite occurs in small foils of brownish yellow color. Magnetite occurs sparingly. Titanite is much more abundant, occurring in large crystals or anhedrons, sometimes wedge-shaped. It includes small feldspar grains, but its crystals also project into larger feldspar grains. The structure of the rock is eugranitic.

An approximate calculation of the analysis of the fresh rock (I) may be made in the following manner: On the basis of a preliminary calculation and estimate there is about 15 per cent of biotite present. For this 1.20 per cent potassa was subtracted. The baryta is calculated as hyalophane, which necessitates a deduction of 0.12 per cent more of potassa. The remainder is calculated as orthoclase. The lime needed for calcite, titanite, and apatite is subtracted from the total lime. The amount of magnetite is estimated. Finally biotite and quartz remain. From the amounts of the bases in this remainder it is estimated that there is 25 per cent free quartz present.

SiO_2	6.51	
Al_2O_3	1.85	
K_2O	1.70	
$KAlSi_3O_8$		10.06
SiO_2	20.78	
Al_2O_3	5.90	
Na_2O	3.57	
$NaAlSi_3O_8$		30.25
SiO_2	6.00	
Al_2O_3	5.09	
CaO	2.79	
$CaAl_2Si_2O_8$		13.88
SiO_2	.59	
Al_2O_3	.25	
BaO	.19	
K_2O	.12	
$Ba\ Al_2Si_2O_8\ 2KAl\ Si_2O_8$		1.15
P_2O_5	.19	
CaO	.25	
Apatite		.44
TiO_2	.66	
SiO_2	.50	
CaO	.49	
Titanite		1.65
FeO	.20	
Fe_2O_3	.41	
Magnetite		.61
CaO	.32	
CO_2	.25	
Calcite		.57
Quartz		25.00

SiO₂	5.85
Al₂O₃	3.85
Fe₂O₃	1.19
FeO	1.71
MgO	1.31
K₂O	1.20
H₂O	.88
Biotite	15.99
Hygroscopic water	.18
Total	99.78

The biotite would have the following composition:

SiO₂	36.59
Al₂O₃	24.08
Fe₂O₃	7.44
FeO	10.69
MgO	8.19
K₂O	7.51
H₂O	5.50
	100.00

The water is somewhat too high, no doubt due to the fact that chlorite and other decomposition products are present.

According to mineralogical composition, habit, and chemical composition, this rock corresponds closely to the granodiorite of the Sierra Nevada, the only difference being that it contains no hornblende, which in the granodiorite is as a rule abundant. It thus occupies a position between a quartz-mica-diorite and a granite. However, as it grades imperceptibly over into more normal granites, it has not been thought worth while to segregate it from that rock. The plagioclase in the Willow Creek rock is, according to the ratio between albite and anorthite, Ab₂An, or a basic oligoclase.

The second rock (80 Boise collection) is grayish white, granular, with a clearly apparent granitic habit. It consists of quartz grains of about the same size as in the fresh rock, while a greenish-gray compact mass replaces the feldspar and has a hardness of about 3. The biotite is replaced by a dull-white micaceous mineral. Small crystals of pyrite abound, and the rock is also traversed by a few small quartz veins. In a few places small grains of zinc blende are seen.

Under the microscope, large grains of quartz with undulous extinction are noted. Between them lies a fine-felted mass of sericite fibers, calcite grains, and in places a little fine-grained quartz. No feldspar remains, though occasionally the outlines of the grains may still be noted. In places larger muscovite foils appear, evidently representing the original biotite. The quartz is in many places attacked by sericitization, fibers and tufts of sericite and calcite developing in it or projecting into it from the surrounding sericite mass. Titanite, extremely abundant in the unaltered rock, is converted into a milky opaque mass, composed of a great number of small crystals. The also abundant apatite is not attacked at all by any metasomatic process. The idiomorphic pyrite is chiefly contained in the altered feldspars, but also included in the quartz. When the pyrite crystals are larger the quartz in the vicinity often shows very strong undulous extinction, but some small perfect crystals also lie in the fresh, unchanged quartz, usually attached at one side to a bunch of sericite fibers. An assay of this rock gave 0.05 ounce of gold and 0.5 ounce of silver to the ton.

Analysis II shows the composition of the altered granite. The chemical change is on the whole slight. Silica shows a small increase, alumina a decrease; a very slight decrease is also noted in the iron oxides, lime, magnesia, and titanic acid, while the whole of the soda and the baryta has been carried away, and the percentage of potash has been considerably augmented. Sulphur and carbon dioxide have been introduced, and the quantity of chemically combined water has been increased.

With the aid of the data furnished by the microscopic investigation, the analysis may be calculated as follows. The lime has been calculated as carbonate, excepting that necessary for apatite and titanite, the whole of the magnesia likewise as carbonate, and as much of the ferrous oxide as the remaining carbon dioxide would permit. The remainder is sericite and quartz. The amount of free quartz has been estimated according to the quantity of oxides available for sericite.

S	.95	
Fe	.83	
Pyrite		1.78
P_2O_5	.17	
CaO	.22	
Apatite		.39
TiO_2	.49	
SiO_2	.33	
CaO	.46	
Titanite		1.28
CaO	2.69	
CO_2	2.11	
Calcium carbonate		4.80
MgO	0.95	
CO_2	1.01	
Magnesium carbonate		1.96
FeO	.90	
CO_2	.55	
Ferrous carbonate		1.45
Quartz		42.00

SiO$_2$	24.33
Al$_2$O$_3$	14.26
Fe$_2$O$_3$	.67
FeO	.44
K$_2$O	4.19
H$_2$O	2.16
Sericite	46.05
Hygroscopic water	.36
Total	100.07

The sericite would have the following composition:

SiO$_2$	52.82
Al$_2$O$_3$	30.96
Fe$_2$O$_3$	1.46
FeO	.96
K$_2$O	9.10
H$_2$O	4.70
	100.00

The altered wall rock from the Black Hornet mine (60, 63 Boise sheet collection) is a grayish-white, very quartzose rock impregnated with small grains and crystals of arsenopyrite. Though adjoining ore contained from $10 to $20, it was found upon assay to contain neither gold nor silver in appreciable quantities. Under the microscope No. 60 appears as a crushed granite or coarse granite-porphyry. The quartz grains are converted into coarse aggregates; the unstriated feldspar is filled with sericite fibers and grains of arsenopyrite, some of which have a little adhering chlorite. No. 63 contains unaltered idiomorphic andalusite, besides crushed large quartz grains and sericitized microperthite, and also a little soda-lime-feldspar. Scattered grains of pyrite are associated with foils of sericite.

The ore of the Black Hornet consists of massive quartz with inclosed sulphides. The preponderating quartz is granular and shows strong evidence of pressure. The ores consist of small, perfect crystals of arsenopyrite and anhedral grains of dark-brown, scarcely translucent zinc blende; inclosed in the latter are many small, distinct foils of sericite. Smaller shreds of this mineral are also scattered through the quartz.

The granite and porphyries adjoining the veins of the Idaho Basin are altered in a similar manner. Sericitization always takes place, and is often accompanied by the formation of carbonates. A specimen from a seam in the hanging wall of the Boulder mine (19 Idaho Basin sheet collection) consists of a quartz vein 2½ inches wide, of which 1 inch is pure quartz and the rest quartz with much finely divided iron pyrite. This vein is inclosed in a greenish-white, bleached granite, impregnated with a little pyrite and arsenopyrite.

An assay of the quartz vein gave 2.25 ounces of gold and 1 ounce of silver to the ton, or a total value of $47.21. An assay of the altered granite immediately adjoining this piece gave only a trace of gold and silver. Under the microscope the altered granite is seen to contain large, partly crushed quartz grains. The feldspar, which is partly orthoclase, partly a soda-lime-feldspar, is filled with sericite foils, chiefly developing on the cleavage planes, and a few grains are almost totally replaced by this material. A few larger muscovite foils probably represent the altered biotite. No calcite is present.

The diorite-porphyrites and quartz-diorite-porphyrites of the Gold Hill and Pioneer mine at Quartzburg show a similar alteration, chiefly consisting in a conversion of the hornblende and biotite to muscovite foils, calcite, and perhaps also pyrite, while there is a great development of fine-felted sericite in the feldspar phenocrysts and in the groundmass. The abundant pyrite is in sharp cubes, usually lined with sericite foils. Aggregates of secondary quartz also develop in places.

Silicification.—Mr. S. F. Emmons, in his studies of the mineral deposits of Colorado and other parts of the Rocky Mountains, has admirably and with deserved emphasis brought out the fact that replacement is a process to which many deposits owe their origin, and that it plays an important part in almost all deposits caused by mineral-bearing waters. Carried away with the importance and interest of these results, many geologists and mining engineers have, however, extended the theory of replacement beyond its proper bounds, and speak of every vein-filling and even of veins of solid white quartz as products of replacement. Against this view a strong protest should be entered. In this connection it may be of interest to consider, briefly, the processes by which silicification may be produced.

In the case of ores consisting of carbonates there may often be some difficulty in deciding what is filling and what is replaced country rock, for carbon dioxide and alkaline carbonates are very strong solvents, attacking easily nearly every one of the rock-forming minerals and forming pseudomorphs after them. The carbonates may replace a rock completely, wholly changing both composition and structure. As an instance may be cited the coarse-grained mixture of carbonates and mariposite (fuchsite—a chromium mica) resulting from the replacement of the serpentine along the Mother lode of California.

As to quartz, the conditions are wholly different. A solution of silica is comparatively inert and does not easily attack any of the rock-forming silicates. Silicification may take place by two greatly differing processes: (1) Cementation, or filling of the interstices of porous or shattered rocks by quartz deposited from solutions; (2) metasomatic silicification, or a substitution of silica for other minerals, the silica either being produced by the alteration of the original minerals or deposited pari passu with the dissolving of the original mineral by active reagents in the waters causing the metasomatic action.

The first process is often observed in the silicification of various sedimentary, porous rocks, chiefly sandstones or tuff or porous igneous rocks, such as certain trachytes and andesites. Silicification by the cementation of shattered rock masses by silica is, of course, a common occurrence in and near quartz veins. But silicification by replacement is a less common process, and is observed chiefly in the case of easily soluble rocks, such as limestone or calcareous shales, when it results in fine-grained or cryptocrystalline aggregates of silica. In the metasomatism of bodies of massive rocks penetrated by chemically active solutions silica is formed in many ways, as by the carbonatization of silicates and sericitization of the feldspars, and if no open spaces are available much of this free silica will be deposited within the rock, usually as fine-grained aggregates more or less mixed with opal and chalcedonite. If no material were added the final result of this would not, however, be a silicification, but merely an increase in the total free quartz of the rock. But in case the rock mass is cut by fissures it appears that most of the resulting free silica is not deposited in the rock, but finds its way out in the open ducts, where, if the solution is supersaturated, it will be deposited. In fact, in the metasomatic processes in the ordinary igneous rocks adjoining gold-quartz veins, late investigations have shown that certain elements are added to the rock, while others, notably silica, are frequently subtracted, to be carried away or deposited in available open spaces.

As for the other possible process of silicification, or a dissolving of the original mineral and a deposition of silica pari passu, it occurs chiefly in easily soluble minerals, such as calcite. In case of the ordinary rock-forming silicates it is apparently not common. The resulting silica is generally in the form of fine, cryptocrystalline aggregates. Rocks silicified by either of these metasomatic processes, or by a combination of both, may occur, but, so far as the writer's experience goes, are not often encountered as wall rocks of auriferous-quartz veins. But neither of these processes can have produced the massive, white, coarse-grained quartz of gold veins belonging to the normal type. This quartz, which contains native gold and sulphides, shows, under the microscope, a peculiar, coarsely granular structure, the grains being partly bordered by crystallographic surfaces. This structure could have been developed only by free crystallization in open spaces. It is scarcely necessary to call attention, in addition, to the frequency of comb structure, etc., proving also the same kind of origin. This does not necessarily mean that all large bodies of quartz have been deposited in an open space, as large as the volume of quartz now is. Repeated openings of the fissure have doubtless often taken place.

In nature the complication of the fissure veins is often great, and it is clear, in fact, that it must be so, for the walls are often shattered, resulting in alteration of the country rock and deposition of a net of quartz in the interstices. Ground-up mud often fills the fissure, and the result of the action of the solutions on this will be a mass of grains of altered rock, cemented as in a sandstone by quartz.

STRUCTURE OF THE VEINS.

The existence of fissure veins is primarily due to one or more fault planes, fissures or seams forming ducts for ore-bearing solutions. The latter have then produced the materials now forming the vein, which may be divided into (1) vein-filling, or minerals deposited in the open spaces along the fissure, and (2) metasomatically altered country rock. Though it is not in every case possible to strictly separate the two classes, as a rule it can be done. Many of the puzzling questions in regard to veins and vein-filling may be solved if this distinction is made and carefully applied. Products of attrition, often present in quartz veins, belong, as a rule, to the second class of materials. The vein-filling which ordinarily constitutes the ore is composed of various sulphides with a gangue of more or less quartz and calcite. Naturally it occurs chiefly along the fault planes and seams. At Willow Creek, for instance, the seams consist of nearly solid sulphides with a little quartz and calcite. These largely represent filling, but are probably, to a minor extent, formed by replacement of the country rock immediately adjoining the fault planes. At other localities, as at Black Hornet and Shaw Mountain, the ore consists of quartz-filling exclusively, with scattered grains and masses of rich sulphides and native gold.

The typical fissure vein may be regarded as a single break or fissure along which, through faulting, more or less continuous open spaces were formed and subsequently filled with ore. On both sides of this filling there is a gradually fading zone of alteration of the country rock.

In many regions the typical simple fissure vein is relatively rare. The country rock may be cut by one or several fault planes, along which only small open spaces have formed and around which there is a wide belt of altered country rock. The ore, then, mainly accumulates along these planes, largely by filling, partly, also, by metasomatism of the adjoining rocks. Again, there may be a shattered zone adjoining one or more fault planes. The rock is then traversed by a complicated system of seams, and large areas of the country rock may be altered. In this case, again, the seams generally contain the gold and the whole seamed rock mass may form a large ore body of low grade.

On Pl. XC (p. 650) a few types of the fissure veins occurring in this region are diagrammatically represented.

MINERAL DEPOSITS OF POST-NEOCENE AGE.

As has been explained before, there is every reason to believe that all of the more important ore deposits antedate the Payette formation, or that, in other words, they are pre-Miocene. But there is also some evidence of a later period of ore deposition, although of less importance, which occurred after the early Neocene eruptions, and which may be going on in depth even at the present time by means of the hot ascending spring waters found at several places in this region.

The Neocene rhyolite area occurring on the Idaho City road 3 miles from Boise is somewhat altered in places and is stated, on reliable authority, to contain $1 per ton in silver and a trace of gold. A Neocene sandstone with veins of opal, occurring near this area along the road, was assayed and found to contain 0.50 ounce of silver per ton. The knob of partly altered augite-andesite near the penitentiary at Boise was assayed and found to contain 0.05 ounce of gold and 0.50 ounce of silver per ton, a total value of $1.38.

The Neocene basalt at the mouth of the canyon of Jackass Creek, Jerusalem Valley, contains a quartz vein which is stated to assay about $4 in gold and silver.

PLATE XC.

649

PLATE XC.

1. Simple fissure vein with one fault plane and with quartz filling. Altered country rock on both sides. Filling only constitutes the ore.
2. Complex fissure vein with three fault planes. Rich ore as filling along narrow openings, partly also by alteration of country rock immediately adjoining fissures. Less altered country rock between and beyond the fault planes.
3. Simple fissure vein without large open spaces. Ore partly as filling, partly as altered country rock. Less altered country rock on both sides of fault plane.
4. Complex fissure vein with two fault planes, along which quartz is deposited as filling. Wide sheeted and altered zone between the two fissures.
5. Irregularly shattered zone between two fault planes. Quartz filling in seams and cracks. Extensive alteration of country rock between fault planes.
6. Single fissure in foot wall of porphyry dike. Dike traversed by stringers from hanging wall, shattered and extensively altered. Quartz filling in fissure and seams. Rich ore in quartz vein. Altered porphyry constitutes low-grade ore.

650

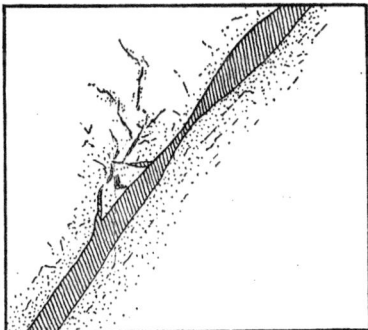

1

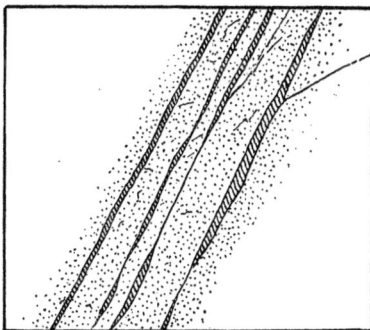

2

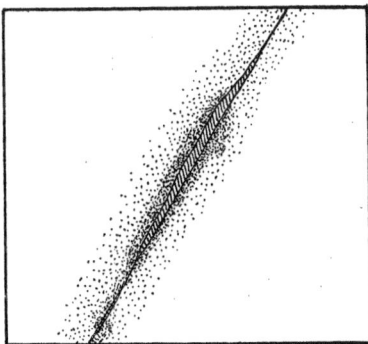

3

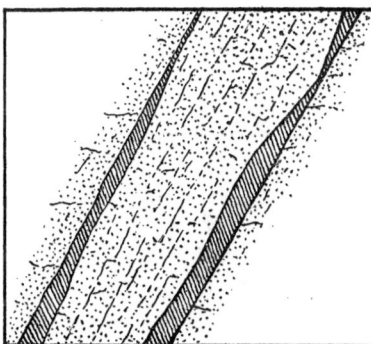

4

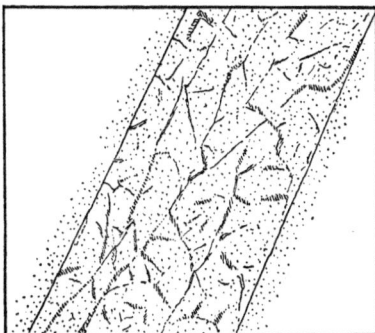

5

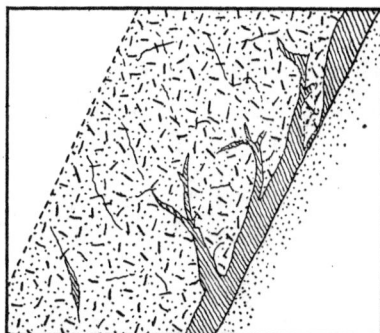

6

Scale.

0 1   10 feet

TYPES OF GOLD-BEARING FISSURE VEINS.

CHAPTER III.

THE IDAHO BASIN.

GEOGRAPHICAL POSITION.

The Idaho Basin includes the headwaters of Moore Creek, a tributary of the Boise River, and is located at a distance of 25 miles in a northwesterly direction from the city of Boise, at about latitude 43° 50' and longitude 115° 50'. In comparision to its size this district has produced a very large amount of gold, chiefly from placer mines, but also considerable from quartz mines. The area embraced in the Idaho Basin—that is to say, the productive part of the same— does not exceed 150 square miles. Its length from north to south is 15 miles and its maximum width about 13 miles.

DISCOVERY AND HISTORY.

The placer mines of the Idaho Basin were discovered in August, 1862, by a party of prospectors from Walla Walla. During the fall of the same year the party is said to have been attacked by Indians, and its leader, Grimes, killed at Grimes Pass. The party, after ascertaining the richness of the placer deposits, returned to Walla Walla and formed a new expedition of 52 men; this party reached the basin the same year. The prospectors located first at the present site of Pioneerville. Subsequently the gold gravels of Centerville were located, and in December of the same year the rich diggings at Idaho City and on Granite Creek were found. Rapidly following these discoveries came explorations of other mineral-bearing parts of Idaho. In 1863 Rocky Bar and the rich mines of Owyhee were found. The influx of miners was extremely rapid after the report of the first discoveries had spread, and one year after the discovery several thousand placer miners were operating in the region. From 1862 to the present date placer mining has been carried on continuously, the operation being limited only by the amount of water available. Naturally, however, the output has gradually decreased since the first years following the discovery, when, as usual in placer regions, the maximum production was reached.

At an early date quartz mines began to be exploited, as it was soon seen that the placer deposits led up to the decomposed croppings of numerous quartz veins. At that time, however, quartz mining was in its infancy, and the ores could not always bear the cost of the treatment, increased in this case by the long distance over which machinery had to be transported and the high wages demanded for labor. At the present time there is a considerable quantity of gravel still left for exploitation, but these deposits will in time be exhausted, and the gold production will then have to depend on the quartz mines. In 1867 and 1868 at least ten mills are reported to have been in operation.

During the last year only a few of the quartz mines and mills were running.

At the present time the Idaho Basin contains five towns, with an aggregate of 1,200 inhabitants. Idaho City is located on Moore Creek, Centerville and Pioneerville on Grimes Creek, and Placerville and Quartzburg on the different branches of Granite Creek.

PRODUCTION.

The total gold production of the basin since its discovery has been the subject of frequent discussion, and it may be said that it is impossible to obtain undisputed data. It is often stated that the production for the first six years amounted to over $40,000,000, and that the total production exceeds $100,000,000. As will be shown, this estimate must be regarded as extravagant. The following table shows the production of the State of Idaho according to the most reliable estimates, contained in the reports of J. Ross Browne and R. W. Raymond, and the later Mint reports:

Production of gold and silver in Idaho, 1863–1876.[1]

Year.	Value.	Remarks.
1863	$7,000,000	Estimate (W. L.). Reports lacking.
1864	6,470,100	J. Ross Browne, report 1867, based on Wells-Fargo data; estimates added.
1865	6,581,400	Do.
1866	8,023,700	Do.[2]
1867	6,500,000	Mint reports.
1868	7,000,000	Do.
1869	7,000,000	Do.
1870	6,000,000	Do.
1871	5,000,000	Do.
1872	2,695,900	$2,300,000 gold, $400,000 silver.
1873	2,500,000	From Eleventh Census, Mineral Industries, p. 40.
1874	1,880,000	Do.
1875	1,750,000	From Eleventh Census, Mineral Industries, p. 40. $1,554,902 in Mint reports.
1876	1,600,000	Reports lacking. Estimate by W. L.
Total.	70,001,100	

[1] During these years the defective statistics do not permit the separation of the gold and silver production, but the latter is relatively small, as shown by the figures for 1872. Few silver mines were worked up to 1876.

[2] $17,000,000 in table on p. 40 of Mineral Industries, Eleventh Census. Authority for this doubtful statement unknown.

Production of gold in Idaho, 1877–1896.

Year.	Value.	Remarks.
1877........	$1,500,000	From table in Eleventh Census, Mineral Industries, p. 41.
1878........	1,150,000	Do.
1879	1,200,000	Do.
1880........	1,980,000	Do.
1881........	1,700,000	Do.
1882........	1,500,000	Do.
1883........	1,400,000	Mint reports.
1884........	1,250,000	Do.
1885......	1,837,400	Do.
1886........	1,798,000	Do.
1887........	2,417,300	Mint reports; $1,900,000 according to Eleventh Census table.
1888........	1,960,000	Mint reports; $2,400,000 according to Eleventh Census table.
1889........	2,055,700	Mint reports.
1890........	1,696,700	Do.
1891........	1,685,600	Do.
1892........	1,721,400	Do.
1893........	1,693,600	Do.
1894........	2,308,800	Do.
1895	2,594,700	Do.
1896........	2,323,700	Do.
Total.	35,772,900	

It appears from the tables that the total production of the State for the first six years scarcely exceeded $41,000,000, according to the Mint statistics, and that consequently the Idaho Basin can not have produced more than a fraction of this amount. Even admitting the more liberal estimates mentioned a few lines below, the total production for this period can not have exceeded $60,000,000. As to the production of the Idaho Basin itself, the data available are still more imperfect. For 1867, 1868, and 1870 there is in Raymond's reports for 1869 and and 1871 an estimation of the total production made by W. A. Atlee, agent of the Wells-Fargo Company at Boise, an authority probably better qualified to judge than anybody else. This includes not only Wells-Fargo shipments, but estimates of all kinds of shipments.

Estimation of the total production of the Idaho Basin in 1867, 1868, and 1870.

District.	1867.	1868.	1870.
Placerville	$363,237	$340,515	$184,428
Centerville	468,556	442,443	249,839
Pioneerville	494,931	552,604	250,000
Idaho City	3,001,568	2,961,213	2,000,584
Total	4,328,292	4,296,775	2,684,851

The same authority estimates a total production in Idaho of $9,000,000 in 1867, $10,000,000 in 1868, and $6,000,000 in 1870. On this Mr. R. W. Raymond comments that his estimate for the total production in Idaho for 1868 is only $7,000,000, and, admitting the excellent facilities which Mr. Atlee had for estimation, still thinks that his values are too high, and is reluctant to allow any increase in his own figures. From 1881 to the present year the statistics are in better condition, and the production of Boise County can be closely estimated. In this production the Idaho Basin furnishes by far the largest quantity of gold; it may be safely said that nine-tenths of the production of Boise County is derived from the Idaho Basin. On the basis of these data I have attempted to construct a table giving the probable production of the Idaho Basin to the present date. It is, of course, only a rough estimate, but it has some value.

During 1884, Plowman's claim, according to the Mint reports, yielded $25,000, Channel's claim at the head of Willow Creek $10,000, and the Granite Creek Company's claim east of Placerville from $15,000 to $20,000. The Chinese companies were during this period estimated to have produced $25,000 a year. In 1883, Placerville, according to the same authority, produced $25,500, Idaho City $381,500, and the Deadwood Basin, located in the northeast portion of Boise County, $8,000. It will be noticed that the greatest production is always that of the Idaho City district.

Table of probable production of the Idaho Basin, 1863–1896.

Year.	Value.	Remarks.
1863	$3,000,000	Estimated (W. L.).
1864	4,000,000	Do.
1865	5,000,000	Do.
1866	5,000,000	Do.
1867	4,300,000	Estimated by W. A. Atlee, in J. Ross Browne's report for 1869.
1868	4,300,000	Do.
1869	3,000,000	Estimated (W. L.).
1870	2,700,000	Estimated by W. A. Atlee (Raymond's report for 1871).
1871	2,000,000	Estimated (W. L.).
1872	1,000,000	Do.
1873	800,000	Do.
1874	700,000	Do.
1875	600,000	Boise County (Raymond's report for 1875).
1876	600,000	Estimated (W. L.).
1877	500,000	Do.
1878	500,000	Do.
1879	400,000	Do.
1880	300,000	Do.
1881	300,000	Boise County, Mint reports.
1882	290,000	Do.
1883	565,000	Do.
1884	400,000	Do.
1885	619,000	Do.
1886	390,900	Do.
1887	502,200	Do.
1888	283,000	Do.
1889	274,600	Do.
1890	320,400	Do.
1891	356,700	Do.
1892	376,400	Do.
1893	280,800	Do.
1894	327,800	Do.
1895	339,000	Do.
1896	326,000	Do.
Total	44,651,800	

It will be seen from this table that the total production amounts to less than $45,000,000. Allowing for the uncertainty in the data, we may safely say that the basin has not produced more than $50,000,000 in gold. Even supposing the admittedly uncertain production of the first four years to be double the amount of estimates here given, the total production would be only $62,000,000.·

While it is not possible to separate the production of the quartz mines from that of the placer mines, it may be said with some confidence that the total production of the former does not exceed $4,000,000.

TOPOGRAPHY.

The Idaho Basin, the topography and geology of which are shown in Pl. XCVI, is located in the middle of that great irregular mountain mass extending between the Salmon and the Snake rivers. Defined more closely, it is situated on the ridge between the north fork of the Boise River and the south fork of the Payette. On the west rises the Boise Range, the summits of which are visible from Boise, and which attains elevations of from 7,000 to 7,500 feet. On the east rise the irregular mountain complexes of Sawtooth Range and its projecting spurs. The basin occupies the head waters of Moore Creek, a tributary joining the Boise River 10 miles above where it leaves its canyon and enters the plains of the Snake River Valley. Ten miles from its junction with the Boise River Moore Creek divides in two, the westerly branch being called Grimes Creek. Five miles above this the narrow canyon in which Moore Creek flows widens out to a broad valley, in which the creek meanders with but little fall. At Idaho City the creek branches again, and both forks head in the high mountains near Wilson Peak and Elk Creek Mountain. A long ridge with a southwesterly direction separates the Moore Creek Basin from the depression of Grimes Creek. This ridge, which for a long distance has a nearly level summit, rises to a height of 1,000 feet above Idaho City. To the south and east of Idaho City the rise is much more rapid. From the vicinity of Thorn Creek Mountain a number of very high and narrow ridges project northward, encircling the southern and eastern part by a chain of hills rising 2,000 feet above Idaho City. Six miles above Idaho City, Moore Creek enters this rugged complex of mountains, and 5 miles farther up heads in the precipitous amphitheaters of Elk Creek Mountain.

The canyon of Grimes Creek reaches to a point 10 miles above its junction with Moore Creek; then the valley broadens, exactly similarly to the valley of Moore Creek, the main branch continuing in a northeasterly direction and heading near Grimes Pass, 4 miles north of Pioneerville. This pass forms the water-parting between the Payette and Boise rivers. It is comparatively low, attaining an elevation of only 5,000 feet. Immediately east of Pioneerville and of Grimes Pass the high ridges of Wilson Peak and Summit Flat rise above the more gently undulating country of the valley of Grimes Creek. An important tributary, Granite Creek, enters 4 miles below Centerville and extends in a northwesterly direction toward Quartzburg. Granite

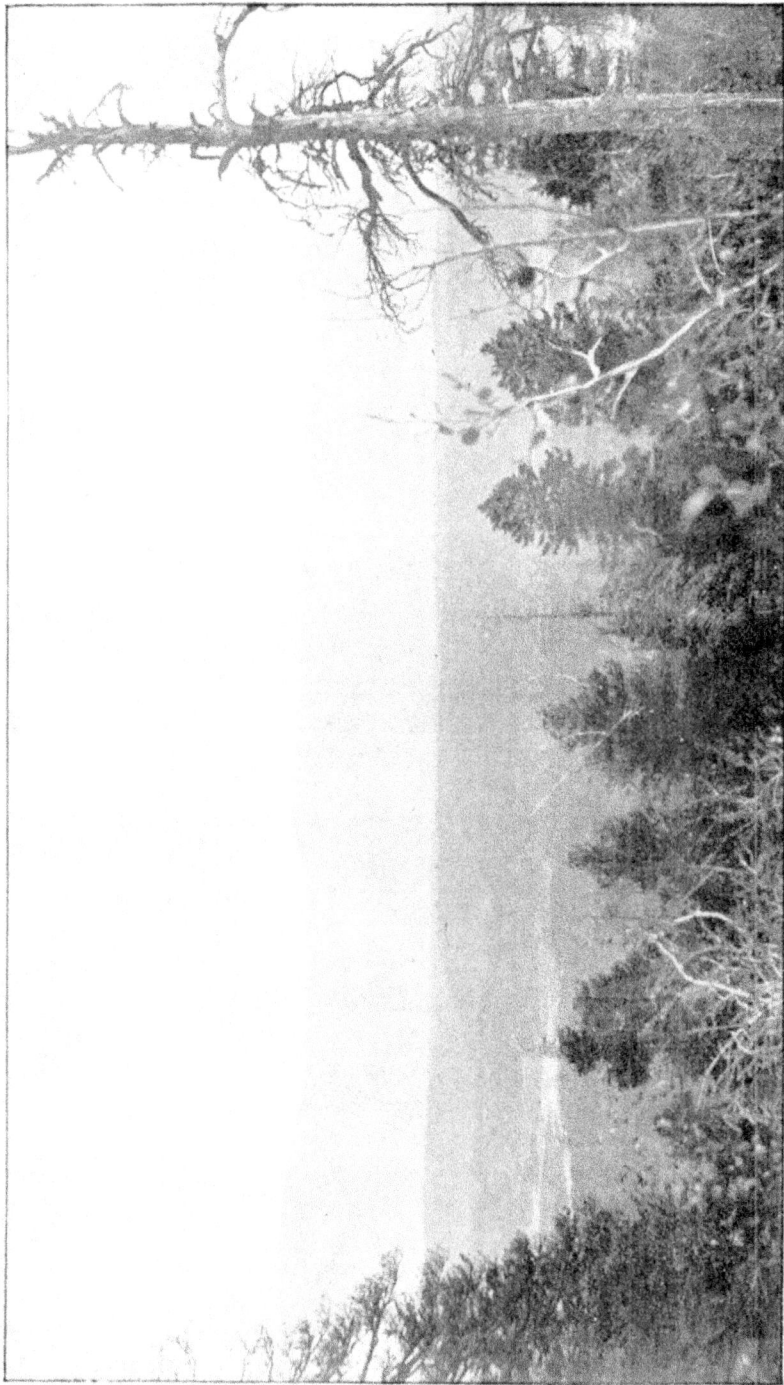

THE IDAHO BASIN, LOOKING EAST FROM THE JERUSALEM ROAD WEST OF QUARTZBURG.

Creek flows in a similar wide valley with gentle slopes, and passes again lead over to the Payette watershed north of Placerville and north of Quartzburg. The lowest pass is the former, which attains an elevation of only 4,700 feet. A short distance west of Granite Creek the Boise Ridge rises very abruptly in the vicinity of Quartzburg, and more gently farther south to elevations ranging up to 7,500 feet. The lowest pass leading across the Boise Ridge, with an elevation of 5,500 feet, is found 6 miles west of Centerville.

One can best appreciate the topographic features of the basin when standing high up on the slopes of the Boise Ridge or on any one of the high peaks rising toward the east. It is seen that the name is extremely appropriate, because it forms a low depression encircled on all sides by a ring of high mountains. Only toward the north, along the water-parting between Boise Ridge and Grimes Pass, is there a lower rim, leading over into the deep canyon of the Payette River. In the accompanying photograph (Pl. XCI), taken from a point at an elevation of 6,000 feet on the road leading from Quartzburg to Jerusalem Valley, the encircling rim from Thorn Creek Mountain to Wilson Peak is well shown, as are the remarkably level ridges separating the lowest depressions in the basin.

GRADES OF THE WATER COURSES.

According to the topographic maps of the United States Geological Survey the grade of the Boise River above and below the mouth of Moore Creek is very slight, being only about 10 feet per mile. Moore Creek for the first 11 miles from its mouth, to the junction with Grimes, has a grade of 40 to 50 feet per mile. The narrow canyon above the mouth of Grimes Creek has a somewhat stronger grade, approximating 66 feet per mile. From Idaho City down to where the canyon begins a grade of 50 feet per mile is obtained; above Idaho City it is for 3 miles 33 feet per mile; beyond this point the grade increases rapidly and is from 100 to 200 feet per mile.

For the first 9 miles from the junction Grimes Creek averages 40 feet per mile; then follow 4 miles of a more abrupt canyon, with a grade of 80 feet per mile. Within the basins of Centerville and Graniteville the creeks have a fall of about 30 feet per mile, which above Pioneerville and Quartzburg rapidly increases to 200 feet per mile.

TERTIARY AND PLEISTOCENE FORMATIONS.

LOWER MOORE CREEK VALLEY.

Configuration of the valley.—From Boise River up to the junction with Grimes Creek, Moore Creek flows in a somewhat broad and U-shaped valley, the slopes of which rise with increasing steepness to ridges with gentle summit lines 2,000 feet above the valley. The hills are covered with a scant vegetation and the soft crumbling granite is easily washed down by atmospheric agencies. The width of the valley between the summits is not more than 2 miles. Above the mouth of Grimes Creek the valley contracts and the steep slopes project, the creek forming a more V shaped valley, which continues to $4\frac{1}{2}$ miles above the junction. A few small alluvial flats lie in the bottom of this canyon.

The basalt flow.—The valley bottom is filled with a basalt flow rarely over one-fourth of a mile in width; its top rises to about 100 feet above the creek level. Through this flow the stream has cut down to a depth somewhat exceeding that which it had attained before the basaltic eruption. The joints in the basalt have produced perpendicular cliffs, the whole being an exact illustration of a canyon within a valley. While much of the basalt flow has been eroded, enough remains to form a nearly continuous fringe of cliffs on either or both sides of the streams, reaching from the mouth of the creek up to $1\frac{1}{4}$ miles above the junction with Grimes Creek. The level top of the flow has been covered by a sloping, sandy wash from the adjoining hills. The character of the valley, the basalt flow and its covering, is well illustrated in Pl. XCII. Above the point mentioned the basalt suddenly ceases, and no more of it, either in outcrops or in pebbles, is seen above. It also extends a few miles up on Grimes Creek, and it is probable that the vent was located at some point in the Grimes Creek Canyon, and that the flow backed up for some distance on Moore Creek. The basalt is about contemporaneous with the flows of the Snake and Boise rivers—geologically speaking recent—being referred to the late Neocene (Pliocene) period.

The present stream gravels.—The bed occupied by the stream is generally narrow, seldom exceeding 100 feet in width. The bed rock is only rarely exposed. The creek is filled with coarse gravel, largely made up of tailings brought down from the Idaho Basin by the winter freshets. The maximum depth of these tailings is 20 feet. The gravels in the present stream have been and are still extensively washed, mostly by Chinese using the ordinary appliances for river mining—dams, Chinese pumps, and derricks. The old channel covered by the recent débris has been exposed by mining at many places, being naturally richer than the tailings. The many basalt bowlders found in it are a considerable obstacle to cheap mining. On Moore Creek, about $4\frac{1}{4}$ miles above the mouth of Grimes Creek, a steam dredge was operated in 1896, with the purpose of striking the rich gravel 10 to 20 feet below the surface. The gravel is reported to have been reached, though the large bowlders interfered somewhat with the work. On the whole, the present stream gravels of lower Moore Creek can not be considered to have been extremely rich.

The gravels below the basalt.—The basalt along Moore Creek is found to rest on stream gravels accumulated in the bed before the molten flow poured down the valley. The old channel is sometimes

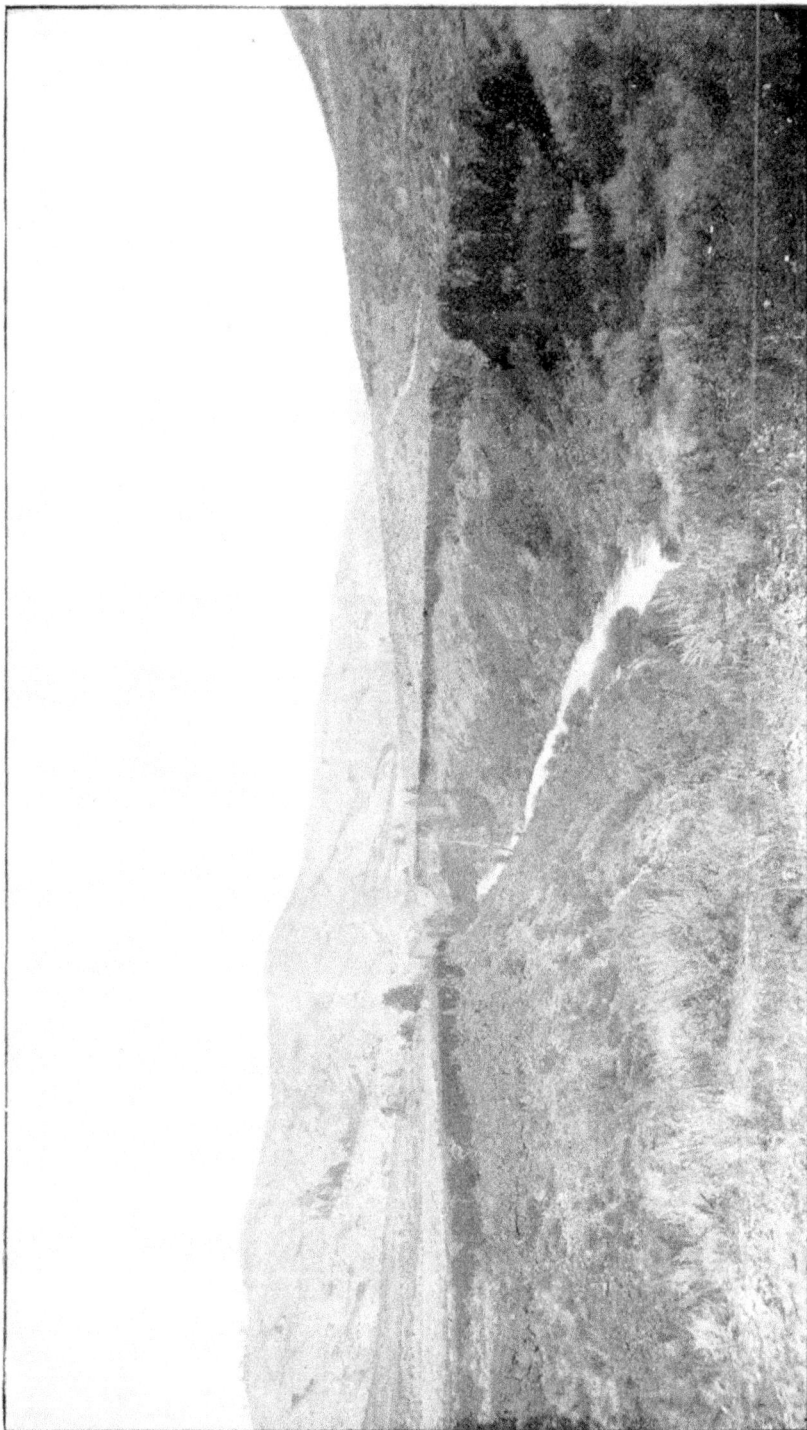

LOWER VALLEY OF MOORE CREEK, 3 MILES ABOVE ITS MOUTH, SHOWING BASALT FLOW CUT THROUGH BY THE CREEK.

preserved on the east side of the creek, sometimes on the west side, which makes it necessary to well establish the course of the channel before commencing extensive work. The old gravel is found at elevations of from 25 to 75 feet above the creek, and has been exposed by a number of tunnels. The gold is fairly coarse, the particles being about the size of mustard seeds, but it can not be said that these old placers have been shown to be very rich. They have not as yet been extensively worked. At a place 1 mile above Half Way House, on the east side of the creek, 6 feet of granitic gravel, covered by 3 feet of sand, is exposed below 30 feet of basalt. One-half mile farther up the old channel is well exposed by two tunnels on the western side of the creek. The old bed rock lies 25 feet above the stream, and the deposit has been developed by 175 feet of tunnels. Above the bed rock lies 8 feet of coarse gravel with a streak of sand in the middle; 70 feet of solid basalt covers this. The gravel is reported to contain about 65 cents per cubic yard. As the gravel is not extremely compact, and as the basalt forms an excellent roof, it might be possible to profitably exploit many of these small stretches of old channels by underground hydraulic operations.

High gravels.—Though the lower part of Moore Creek has not been thoroughly examined, it has been shown that gravels exist on some of the high ridges within this drainage. The study of these high gravels, which are probably of Tertiary age, is of the highest importance, as it enables us to trace more fully the geological history of the region. The largest of these high gravels was found on the summit of a high and narrow ridge between the two forks of Thorn Creek, at an elevation of 4,500 feet. This deposit, which consists of extremely well-washed gravel of granitic and quartzose character, has been worked to some extent by the hydraulic process and found to contain gold in paying quantities. The extent from east to west is only a few hundred feet. The depth of gravel is probably considerable, although difficult to estimate on account of slides on the steep slopes. Small patches of gravel are also reported to exist on the neighboring ridges. It is probable that during the highest stand of the Neocene lake which occupied the whole lower basin of the Snake River gravel deposits filled the upper parts of Moore Creek and the Boise River. The occurrence of this high gravel will be discussed again later in connection with the geological history of the Idaho Basin.

UPPER MOORE CREEK VALLEY.

Configuration of the valley.—Five miles below Idaho City the form of the valley suddenly changes. From a level bottom the slopes rise gradually, numerous creeks branch, and the whole forms a broad, basin-like depression. The general character is well shown in Pl. XCIII, looking southwest from Idaho City. Three miles above Idaho City a narrow canyon begins again. Seen from some elevation, the three most impressive topographic features are (1) the broad valley; (2) the level ridges between Idaho City and Centerville, between Elk Creek and Moore Creek, and between Granite and Bannock gulches, rising to an elevation of 1,000 feet above the valley; and (3) the encircling rim of high, deeply dissected peaks and ridges, attaining a height of 2,000 or 3,000 feet above the valley.

The present stream gravels.—The alluvial gravels filling the bed of the present stream form broad flats, over which the water course meanders in changing channels. The largest part of these gravels is débris from the hydraulic mining operations carried on in the bench gravels and the high gravels. The width of the tailings below Idaho City reaches 1,000 feet. In two places the channel contracts between low, projecting hills. At Idaho City the maximum width is reached; here the tailings are about 2,000 feet wide, contracting again to a narrow channel 1½ miles above the city. Two and one-half miles above the same place a comparatively narrow canyon begins, and the tailings are only up to 200 feet wide. The tailings cover the imperfectly washed original creek bed, and in many places also the first (lowest) terrace or bench. At Warm Springs they are reported to be 15 feet deep; at Idaho City as much as 40 feet. They consist chiefly of granitic and porphyritic pebbles, with much sand. By a natural process the gold in the tailings is gradually concentrated. At some time the whole creek bed will probably be washed over again to recover this gold and to reach the older bottom gravels, which are supposed to be rich in many places. Probably the only way in which this can be profitably done is by means of hydraulic elevators or dredges.

The stream gravels were most extensively worked in early days. Moore Creek was very rich up to the mouth of Gambrinus Gulch, though gravels have been washed still higher up. The gulches entering from the south, as a rule, paid only for a short distance from the main creek; their upper courses were nearly barren. On the northern side, Gambrinus and Illinois gulches were extremely rich. Bear Gulch has also been worked extensively. The bottom of Elk Creek in its upper course is generally narrow, but was rich up to Boulder mine; 2 miles below Boulder mine the creek bottom widens to 200 feet, and placer diggings were being operated in 1896. Deer Creek, heading at Summit mine, yielded very heavily, the output of the half mile near the summit being placed at $90,000. Wolf Creek has been worked, but was not so rich. Spanish Creek was also worked, and near its head lies a flat, one-eighth mile wide and one-half mile long, covered by 12 feet of angular wash, which has been extensively worked, the gold being probably derived from small seams or from a gold-quartz vein not yet discovered.

The gulches entering from the north and south below Idaho City were generally barren beyond the extent of the terraces.

Bench gravels.—Where the broad valley opens, 6 miles below Idaho

IDAHO CITY, FROM GOLD HILL, LOOKING SOUTHWEST, SHOWING TAILINGS AND BENCH GRAVELS.

City, a series of shelf-like terraces, entirely absent farther down, begin to appear. The width is only a few hundred feet, and rarely are longer stretches of them preserved, each little creek and gully usually cutting the terrace in two. For the first 4 miles most of the gravel patches lie on the northern side of the valley. At least two terraces may here be recognized, the bed rock of the lower one being 50 and of the upper about 100 feet above the creek. The depth of the gravel is seldom over 30 feet, though near the bed-rock slopes much débris has slid down over the gravel. All of these gravel patches have been very rich, and work is still in progress on some of them. The bed rock, as far up as 1 mile below Warm Springs, is granite or granite-porphyry.

Above Warm Springs the gravel terraces or benches are very pronounced and often form continuous streaks of considerable length. The best exposures are found at Turner's claim, 1 mile below Idaho City, on the road to Warm Springs. At this place the gravel terraces occupy a total width of one-half mile, and rest partly on gran-

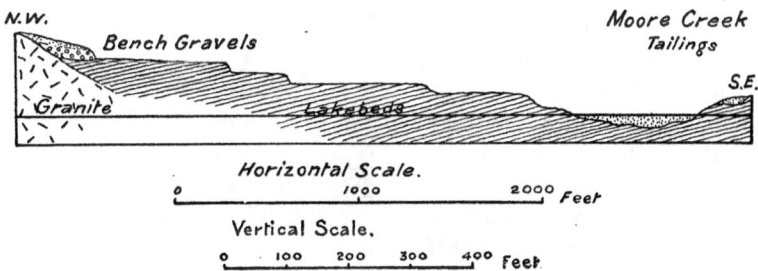

FIG. 55.—Gravel benches 1¼ miles below Idaho City.

ite, partly on Tertiary lake beds, gently inclined westward. Fig. 55 shows a profile of the different terraces here exposed. The highest terrace is found at an elevation of 80 feet above the creek bed. Below this there are three others, and possibly four, at intervals of 15 and 30 feet. The lowest terrace is said to be covered by the tailings. These gravels have been very extensively washed, but a considerable amount still remains. The upper terrace, illustrated in fig. 56, is covered by 8 to 12 feet of well-washed gravel, chiefly granitic in character. This contains the largest part of the gold, and rests on the eroded surface of the soft lake beds. This pay gravel is again covered by 12 feet of fine sandy or clayey sediments with occasional carbonaceous seams. In this there is but little gold. On top rests 8 feet of angular surface gravel, washed down from the adjoining hillside. This gravel is barren. The gravel terrace extends in a northeasterly direction along Elk Creek, and is chiefly developed on the western side. The continuous bodies give way to isolated patches, and 2 miles above Idaho City Elk Creek passes into a narrow canyon. In Idaho City gravel terraces are noted surrounding a high body of older gravel called East Hill. A part of this terrace, on which the town stands, has not yet been mined. Opposite Idaho City, and in fact all along the southern side of Moore Creek up to the mouth of Granite Creek, the gravel benches are more or less continuous. The lowest is noted 20 feet above the creek and higher ones are at 50 and 100 feet. Small benches occur on the north side of Moore Creek east of Idaho City. At least three different benches may be recognized at the elevations noted. Most of these bench gravels near Idaho City rest on lake beds of sandy or gravelly character, usually referred to by the miners as "false bed rock."

All the small patches of gravel terraces have been extensively washed and are not yet quite exhausted. The thickness of the gravel rarely exceeds 25 feet.

Above the mouth of Granite Creek, Moore Creek enters a canyon, along which there are but small indications of gravel terraces. At Plowman's sawmill the creek widens somewhat, and from here up the bottom is generally occupied by a flat 100 to 300 feet wide, the surface of which is 10 to 20 feet above the creek. Upon the sides are occasional benches at an approximate height of 50 feet above the creek. All these low terraces have been washed as high up as the mouth of Gambrinus Gulch, and from the evidence of the old washings it is clear that most of the gold came down from this gulch. Above

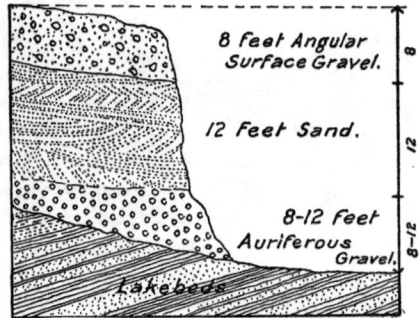

FIG. 56.—Section of highest bench, 1¼ miles below Idaho City.

Gambrinus Gulch the washings are less extensive, and some of the low gravel flats have never been worked, being evidently too poor. Two miles above Idaho City, as stated above, Elk Creek enters a narrow canyon, and although this widens somewhat farther up, still there are only occasional patches of gravel terraces remaining, at elevations of about 50 feet above the creek. The manner of mining the bench gravels by the hydraulic process is shown on Pl. XCIV.

High gravels.—There are several bodies of gravel in the vicinity of Idaho City which in their occurrence differ from the ordinary bench gravels, and which are generally at high elevations above the creeks. While these also are probably former terraces of the valley of Moore Creek, they are much older, and it seems desirable to treat them separately.

The slope opposite Idaho City is occupied up to 400 feet above the creek by soft lake beds cut up into sharp ridges separated by deep and narrow ravines. The gently sloping tops of these ridges are

WORKING BENCH GRAVEL BY THE HYDRAULIC PROCESS AT IDAHO CITY.

covered by auriferous gravels which have
a thickness of 60 feet or less. The base
of these gravels lies, at Barker's claim, 300
feet above Moore Creek. This gravel shows
excellent fluviatile stratification, and appears
to rest in a flat channel eroded in the lake
beds. The gravel consists of well-washed
granite pebbles, accompanied by some of
quartz. Bowlders 2 feet in diameter occur
occasionally. This gravel is quite rich, the
best pay being found, as usual, on the bed
rock, i. e., on the soft lake beds, but there is
also gold distributed through the gravel.
The gravel and gold has also, of course, slid
down the steep hillside, and this material
covering the slopes has been washed. During the summer of 1896 the claim of Mr.
Barker, 1 mile southeast of Idaho City, was
worked by the hydraulic process. The pay
gravel does not extend southward beyond
the lake beds. In these high gravels there
occur somewhat abundant large cobbles up
to 1 foot in diameter of quartz stained brownish, and which are often rich in gold, one
bowlder sometimes yielding $20. The source
of this quartz is not known. It certainly does
not come from the high hills to the south.
Fig. 57 shows the general relations of the
high gravels at Idaho City.

Another important body of the high gravels
is found on the hill immediately east of Idaho
City. The gravel, which reaches a total depth
of 100 feet, forms a body about 2,000 feet long
and 1,000 feet wide. It rests throughout on
clayey and sandy lake beds, dipping gently
westward at an angle of 10°. Seen from the
south side of the creek, the stratification
planes in the gravel appear to have a decided
dip westward, amounting to 4°, or a little less
than the underlying rocks. The gravel rests
remarkably evenly on the lake beds, with
but little sign of unconformity. The geological section in fig. 57 illustrates the occurrence, while Pl. XCV shows the gravel bank
and underlying lake bed in Plowman's claim.
A marked fluviatile stratification is often

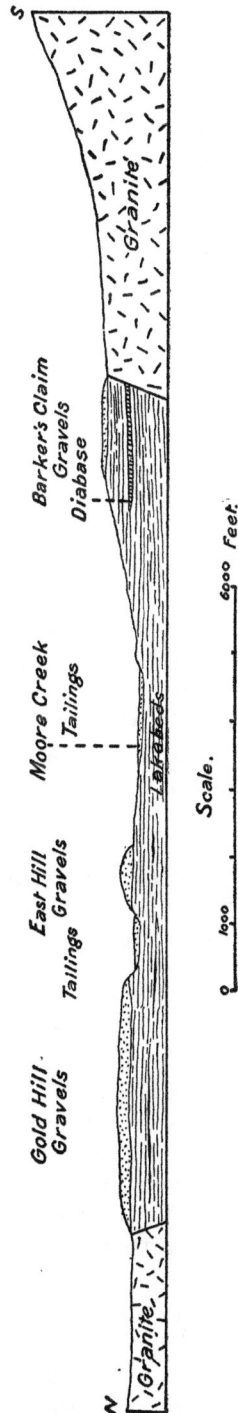

FIG. 57.—Section across lake beds and gravels at Idaho City.

visible. The gravel is medium coarse and is made up chiefly of granitic bowlders with occasional pebbles of quartz. By tracing this area around, it is found that the bed rock rises gently, being at the eastern end at least 200 feet above the creek. At the western end the exposures are not so good, as tailings and bench gravels lie up against the higher gravels and the underlying beds are not here visible. The gravel on East Hill, as this area is called, has been worked by the hydraulic process for a number of years, and the claim is reported to have produced from $10,000 to $20,000 a year. The largest part of the gold is found resting on the soft bed rock. The upper part of the gravel also contains some gold, but probably not more than about 5 cents per cubic yard. The gold is fairly coarse, and has a value of $16.50 an ounce before melting. On the surface of the gravel the gold appears to be more abundant, which is probably caused by a gradual concentration by atmospheric agencies.

Another large body of gravel is that known as Gold Hill (fig. 57), occupying about 160 acres and situated on the point between the creek and Bear Gulch. The top of this gravel body is 350 feet above Idaho City, and its greatest depth is probably not less than 200 feet. Along Elk Creek and Bear Gulch the gravel rests on the same soft lake beds which crop out on East Hill. Toward the southwest tailings lie up against it. On the northeast side, on the other hand, the gravel rests directly on granite bed rock. Here also the dip of the gravel beds is about 4°, while that of the lake beds is 10° W. Good exposures, made by hydraulic mining, are seen on the southwest side. One-fourth mile north of Idaho City the surface of the soft lake beds lies 20 feet above the creek. Above the lake beds lies 15 feet of coarse heavy gravel with many subangular fragments. Above this is 10 feet of sand with clay streaks, sometimes a little coaly, and somewhat resembling the lake beds. Capping this sand is ordinary well-washed gravel. Here, as in the other gravel bodies, the largest pay was found on the surface of the bed rock—that is, on the surface of the lake beds—and this rich bottom stratum has been mined by the drifting process both on the southeast and on the southwest side. A large body of gravel here remains, which can be worked by the hydraulic process, although it is probably of low grade. Owing to some difficulty in procuring water, the claims on this hill have not yet been extensively worked.

On the ridges above these deposits no other gravel masses have been found, but a small body rests on a sidehill to the east of Elk Creek, 3 miles northeast of Idaho City, at an elevation of about 200 feet above the creek. On the western side of Elk Creek, opposite Idaho City, small patches of gravel are occasionally found 200 feet above the creek level, and scattered pebbles occur in many places at about this elevation. Still another body of high gravel is that found on the point between Granite Creek and Moore Creek, east of Idaho City.

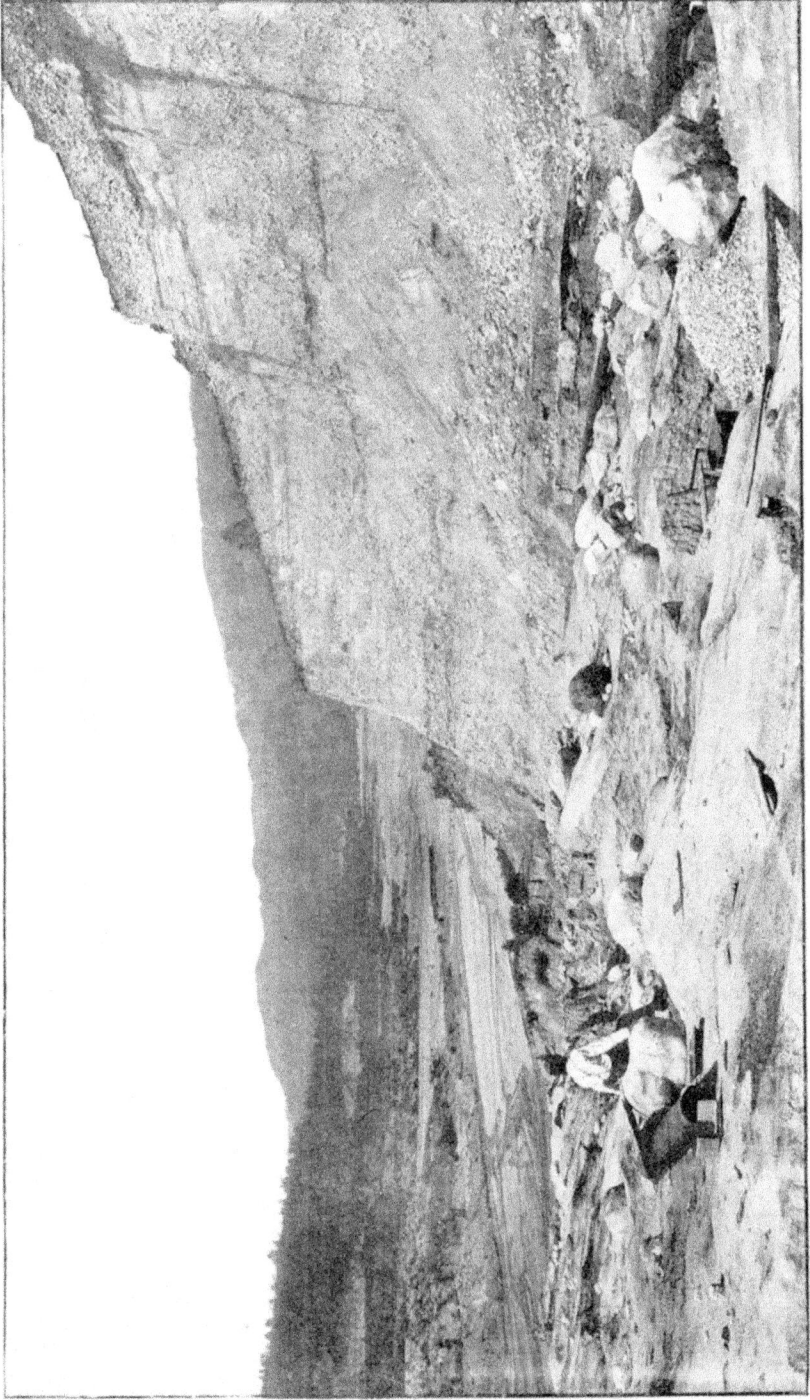

GRAVEL BANK WITH UNDERLYING LAKE BED AT PLOWMANS CLAIM, EAST HILL, IDAHO CITY, LOOKING SOUTHWEST.

It forms a bench-like deposit 175 feet above the creek, and the larger part of it has been worked by the hydraulic process.

Lake beds.—Near Idaho City, on both sides of Moore Creek, a considerable area is covered by soft sedimentary beds of clay, sand, and gravel. These do not bear evidence of being river deposits. The regular stratification and their general character indicate that they have been deposited in a body of water, in all probability a freshwater lake. These beds do not contain any notable amount of gold, but the fluviatile auriferous gravels just described are deposited on their eroded surface. The dip of the beds proves that that they have been disturbed since their deposition. Recognizing their sedimentary nature, the miners call these lake beds "false bed rock." The lake beds are first met with 1 mile below Warm Springs, where they appear as semiconsolidated white sand and clays. Near Warm Springs they assume the form of coarse yellowish sandstones, evidently cemented by the opaline silica of the hot waters. In fact, sandstone cemented

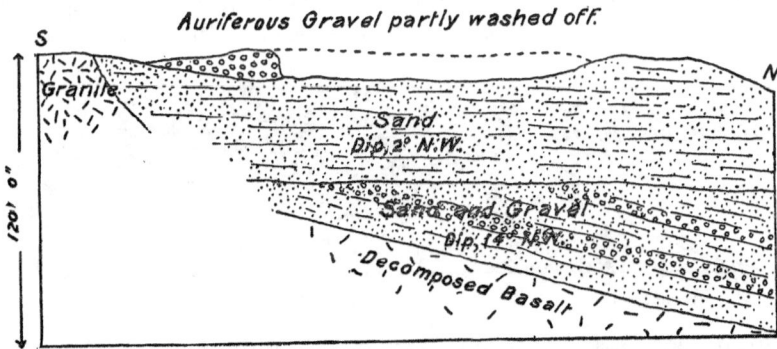

FIG. 58.—Exposure of lake beds and auriferous gravel 1¼ miles south of Idaho City. Top of gravel, 400 feet above Idaho City.

by fire opal has been collected on both sides of the creek, and the occurrence is mentioned by Mr. S. F. Emmons.[1] Owing to the very coarse character and rapid accumulation of the débris the stratification is poorly preserved. Half a mile below the hot springs these sandstones form a bluff 300 feet high, and also extend east of the road, producing a relatively narrow passage for the creek. A slide has taken place near the creek, covering an older channel, the bed rock of which lies only a few feet above the present creek and which has been mined by several tunnels. At Turner's claim the lake beds form the bed rock and are well exposed. They dip 5° NW. and consist of soft sand and clay with occasional coaly layers. The clay encloses nodules of iron pyrite containing some gold and silver.

Near Turner's claim a projecting spur of granite reaches to within a few hundred feet of the road, and in the hydraulic cuts the lake beds are seen to rest on it. Between Turner's claim and the city the

---

[1] Boise Statesman, March 29, 1896.

lake beds are not exposed, but the gravel of the terraces is reported
to rest on them. Across Elk Creek from Gold Hill sandy and clayey
lake beds again appear, the bench gravels resting on their eroded
surface. Granite appears a short distance up on the hillside.

Around East Hill and Gold Hill the lake beds are well exposed, and
consist largely of greenish or gray clay with arenaceous streaks and
intercalated beds of black clay with coaly streaks. The dip is 10° W.
A few fossil plants were found here, which were identified by Mr.
Knowlton. (See Appendix, p. 721.)

These plants identify the lake beds with the Payette formation of
the foothills, a correlation which the field work had shown to be very
probable. Considerable masses of fossil wood are reported to have
been found in the lake beds at the mouth of Steamboat Gulch, 1 mile
southeast of Idaho City.

The lake beds attain their greatest development south of Idaho City,
where they are more than 300 feet thick. Their character is here
prevailingly sandy, with some medium-coarse gravels and with occasional coaly and clayey layers. The exposure illustrated in fig. 58

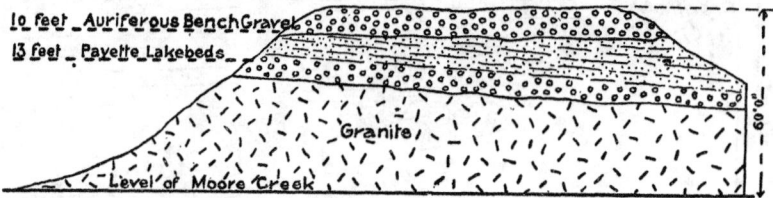

FIG. 59.—Bench gravel and lake beds at mouth of Granite Creek, 2 miles west of Idaho City.

indicates an apparent unconformity in the lake beds. They continue
up the creek, decreasing in width, to the mouth of Granite Creek.
Good exposures are seen at Brockhausen and Spiro's claim, between
Granite and Bannock creeks, where the bench gravels rest on them.
Fine gravels here appear in the lake beds, interstratified with clay and
sand. At the little bench just south of the mouth of Granite Creek,
12 feet of lake beds, dipping 4° W., at first gravelly, then sandy, rest
on granite, and on the eroded surface of the lake beds rests a patch
of bench gravel 50 feet above the main creek. A few quartz pebbles
are present in the gravel of the lake beds (fig. 59).

A small area of clayey lake beds is said to exist on the high plateau
several hundred feet above Moore Creek and about 2 miles east-
southeast of the mouth of Granite Creek.

The contacts of the lake beds with the granite offer points of great
interest. It has already been noted that lake beds rest on granite
near Turner's claim, at the east end of Gold Hill, and at the mouth of
Granite Creek. Everything indicates that they were laid down on an
uneven surface, and probably in a valley with configuration similar
to that of the present Moore Creek Basin. But at many other places

it is clear that the lake beds are separated from the granite by normal faults. One mile below Warm Springs, on both sides of Moore Creek, there is evidence that the lake beds abut directly against the granite. The same relation is noted at the sandstone bluff, 300 feet high, back of Warm Springs. Here the almost horizontal lake beds abut against a steep granite bluff, and a little lateral valley has formed along the contact. There can hardly be any doubt that this steep bluff represents a fault-scarp. The deeply incised gulch just southward gives similar testimony as to the sharp abutment of the two formations against each other.

A small hot spring is located at a point on this fault, and the large warm springs probably also issue from this fault, though at present they break out through the sandstone a little below it. The water has a very high temperature, and the total quantity is probably not less than 100 miner's inches. The water is not rich in dissolved salts, but has a slight smell of sulphureted hydrogen. Mr. J. B. Hastings[1] thinks that this fault and the accompanying slipping down of this mass of conglomerate or sandstone caused the damming of a lake and the deposition of the lake beds. This can not be accepted as a correct explanation, for the sandstone is of the same age as the lake beds, the deposition of which was caused by events much farther reaching than a landslide.

All along the southern contact line of lake beds and granite, from Moore Creek below Warm Springs to beyond Bannock Creek, the evidence of a fault is very decided. Nearly everywhere along this line the lake beds cease suddenly, and south of them the granite rises in a steep bluff, contrasting strongly with the confused topography of the soft and sliding lake beds. The best evidence is found at the contact back of Barker's claim, where the contact plane between granite and lake beds is found to dip at an angle of 45° N. A similar and extremely well-exposed fault is shown on both sides of Elk Creek at the northern end of Gold Hill at the mouth of Lincoln Gulch.

In conclusion, the lake beds of the Payette formation in the Moore Creek basin form an area of about 7 square miles; they dip west or northwest at angles of from 4° to 14°, and they are often separated from the granite by marginal faults. They are probably a remnant of a more extensive area preserved by reason of being sunk in the granite by movement along these faults.

A total thickness of from 300 to 400 feet is exposed above Moore Creek. The idea that rich gold gravel would be found below the lake beds led in 1894 to the sinking of a shaft and bore hole at Idaho City, only about 20 feet above the level of Moore Creek. One hundred feet of shaft were sunk, and then, when the water became too difficult to handle, a bore hole continued down to a total depth of 516 feet. It is reported that granitic bed rock was struck at that depth. Samples

---

[1] Eng. and Min. Jour., July 21, 1894.

of boring show mainly granitic sand and clay, with nodules of iron pyrite occurring at frequent intervals. The total thickness of the lake beds would, then, be not less than 850 feet. A few miners'-inches of saline water flowed from the well, as might indeed be expected, for the geological conditions are such that a typical artesian basin is formed.

Gold in the lake beds.—It is stated that no gold is contained in the lake beds or the false bed rock. This certainly seems, at first glance, to be a strange state of affairs, considering that the sands, clays, and gravels of the lake beds consist of practically the same material as the auriferous gravels. It would seem to imply that the quartz veins from which the gold was derived were formed between the period of the lake beds and that of the gravels. It will be shown, however, that the lake beds are not entirely void of gold. Those just south of the mouth of Granite Creek, shown in fig. 59, were prospected with the pan. In the lowest bed, consisting of coarse gravel with much sand, nothing was found; but 6 feet above the granite, in a finer gravel admixed with some quartz pebbles, several colors were found in every pan. The gold is extremely fine and of a rather pale color. There is very little black sand in this gravel, but a considerable quantity of monazite. The samples were taken under conditions that made it impossible for any of this gold to have been derived from the rich gravel above. Mr. Brockhausen informed me that in a claim one-half mile below the mouth of Granite Creek a considerable amount of gold was taken out of a bed of gravel dipping below the false bed rock.

At the mouth of Noble Gulch, opposite Idaho City, there is a bed of gravel a few feet thick dipping below strata of carbonaceous clay. This gravel has been worked and is reported to have yielded some gold. Mr. Barker informs me that a little gold may occur in the gravels of the lake beds wherever quartz pebbles are present. Mr. Kramer, who owns a claim one-half mile below Warm Springs, states that John Wood, former owner of the claim, obtained good prospects in a bed of gravel dipping under the false bed rock at that place. The locality is now covered up. Mr. Turner states that nodules of pyrite, containing a few dollars in gold and silver, are often found in the false bed rock. During the sinking of the artesian well at Idaho City certain strata were found to contain much iron pyrite, which upon being washed out and assayed was found to carry as much as $12 per ton in gold and silver.

It is conceded, however, that these occurrences of pyrite do not necessarily indicate an original content of gold in the lake beds, as the precious metals may have been leached from the overlying rich gravels and deposited with the pyrite below.

It is thus certain that free gold occurs in some of the gravel of the lake beds. That there could not be much of it present is clear from the mode of formation of the lake beds, for they were deposited by rapid accumulation in a body of water affording no opportunity for concentration. Moreover, the detritus was mostly derived from the immediately surrounding hills, which are nearly barren of mineral veins, while the overlying gravels were transported, concentrated, and assorted by streams coming from the region of auriferous quartz veins. It is probable that a system of streams existed before the lake beds were laid down. It would naturally be expected that many of these should carry gold, and it is indeed probable that if the lake beds were removed we should find auriferous gravels on the bed rock along these lines of old stream courses. But it is extremely unlikely that a random bore hole would strike any of these deposits. They could be found only by prospecting along the margin of the lake beds and following down any rich stratum that might be found. In this case, however, the cost and difficulties of mining would be very great.

Olivine-basalt (dolerite).—One mile above Idaho City, on the north side of Moore Creek, there are peculiar outcrops of a black or dark-green very tough rock, weathering in rounded outcrops, which are commonly referred to as "nigger heads." Decomposing, they yield a dark-red, yellow, or green clayey soil. Gravel benches rest upon this rock, which apparently forms an intercalated bed up to 100 feet thick in the lake beds. The sheet lies flat at the point indicated, crosses Moore Creek in a narrow strait 2 miles above the city, and then, probably being tilted, rises to elevations of 400 feet above the creek near Pine Gulch. Here it is evidently separated by a fault from the granite. The same rock is found again in nearly every one of the small gulches entering Moore Creek opposite Idaho City, and here it is covered by a considerable thickness of lake beds and is exposed only in the bottom. It is a medium-grained, dark-green rock, with abundant scattered crystals of greenish-yellow olivine. Under the microscope it is seen to be a coarse olivine-basalt (dolerite), and to consist of large phenocrysts of olivine and small crystals of magnetite as the earliest product of consolidation. There is, further, a large amount of violet-brown augite in large anhedral individuals, forming a sort of base, in which are embedded the irregularly distributed laths of a basic feldspar (labradorite). The olivine decomposes to brownish-red products, also to serpentine; the augite to chlorite, with beautiful radial structure. The structure of the rock is really that of a diabase. This rock was evidently poured out on the surface as a lava at the time of deposition of the lake beds.

THE VALLEY OF GRIMES CREEK.

Configuration.—Four miles below Centerville the canyon widens to a broad valley with gentle slopes, similar to that of Moore Creek, extending in an east-northeasterly direction for 11 miles. Above this Grimes Creek makes a sudden bend, and, separated only by a low ridge from the deep canyon of the Payette River, finally reaches its headwaters at Summit Flat. Only two tributaries join it—Clear Creek, heading in the rugged mountains near Wilson Peak, and Muddy Creek, heading a few miles northwest of Pioneerville.

Present stream gravels.—A large amount of tailings lies nearly all along Grimes Creek. At Centerville they are 900 feet wide; farther up they narrow considerably, where the hills approach closer to the creek, to widen again near Pioneerville; above this place the creek enters a rather narrow canyon. A great mass of tailings also lies in Muddy Creek. Clear Creek has never been washed, and the original wide alluvial grassy flats are here preserved in the lower course of the creek. Concerning the gold content of these tailings, the same remarks apply here which were made in relation to those of Moore Creek. They can doubtless be worked profitably in many places by means of hydraulic elevators or dredges. The difficulty is to obtain a sufficient water supply. The beds of Grimes Creek and Muddy Creek are reported to have been rich throughout, while Clear Creek did not pay well. Only a few of the side gulches contained gold in paying quantities. Willow Creek was rich, and contained near its head a body of angular gravel, known as Channel's claim, which has yielded much gold. Henry Creek, leading up to Summit mine, was also rich. The bulk of the gold appears to have come down from the headwaters of Grimes and Muddy Creeks.

The tailings are very sandy, being composed of almost 60 per cent sand and 40 per cent cobbles, and the maximum depth is 15 to 20 feet. The lower part of the tailings and benches of Grimes Creek, up to 3 miles below Centerville, is owned by the Grimes Creek Bed Rock Flume Company; the upper part, as far as several miles above Pioneerville, by the Wilson Company, which for many years has carried on active operations near Pioneerville. The creek has been worked as far up as the big bend at Grimes Pass, where the gravel is characterized by a great many heavy bowlders of porphyrite. Many similar cobbles of porphyries also occur farther down on Muddy and Grimes creeks. Pebbles of obsidian have been found on Muddy Creek.

Bench gravels.—As along Moore Creek, gravel benches occur at different elevations all along Grimes Creek. The two most prominent benches are at elevations of 30 and 60 feet above the present creek bed, but scattered gravel occurs at higher elevations also. Such is the thin gravel occurring near Centerville up to 150 feet above the creek. These bench gravels have been very extensively worked, and but little remains of them near Centerville. A short distance above Centerville, at a place called Bummer Hill, they were of unusual richness. Above the narrow canyon the bench gravels appear again in the open valley; where Muddy, Clear, and Grimes creeks join the same kind of benches are noted and hydraulic work has been actively prosecuted. Pioneerville is situated on the lower bench, 25 feet above the creek. A low bench gravel, one-half mile above town, worked in 1896, is illustrated in fig. 60. The bed rock is at the creek level, and the work was done by means of hydraulic elevators, which have been extensively utilized by this company. The gold, which is generally of the size of mustard seeds, lies often on the higher bed rock instead of in the potholes. The lower gravel carries all the gold.

Older gravels.—The low terrace separating Muddy and Grimes creeks, one-fourth mile northwest of Pioneerville, and covered by later bench gravels, is partly made up of granite, partly of an older, somewhat cemented, granitic gravel. This belt is only about one-fourth mile wide, and the gravel abuts against the granite on the northern side, showing that it is separated from it by a fault. It is evidently an older gravel sunk down along a fault-line. On the southern side it apparently rests on granite. The same fault is well shown in the creek on the eastern side, one-fourth mile above the town. It is reported that a shaft 100 feet deep was sunk in this gravel some twenty years ago, without reaching the underlying granite. The gravel contains some coarse gold. The side hill immediately east of this fault is wholly composed of granite, but above, on the narrow ridge separating Grimes and Clear creeks, lies a body of somewhat similar gravel, attaining an elevation of 200 feet above the creek. Gran-

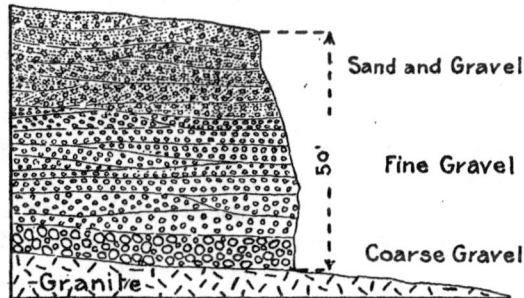

FIG. 60.—Bank of bench gravels one-fourth mile north of Pioneerville, level of Grimes Creek.

ite bed rock is found 120 feet above the creek, and on it rests a bed of coarse gravel containing some gold. Much of this gravel has been worked by the hydraulic process.

Lake beds.—No extensive areas of lake beds occur in this vicinity, but in the valley of Muddy Creek, 2 miles north of Pioneerville, a small patch of a rather remarkable deposit is found. For a distance of a mile the creek runs in a narrow canyon, which then opens to a somewhat wider valley. Here lie, chiefly along the western side of the creek, beds of soft white sands, gravels, and clays, with a little lignite. In one place where the stratification could be made out the layers seemed to dip 20° W. This small mass of sedimentary deposit has a lacustrine character, and should probably be correlated with the lake beds of Idaho City; it does not contain any gold, but the surface gravels resting on it are said to have been unusually rich, the softer beds having acted as riffles, catching the gold. On all sides the granite rises rather steeply. At Bummer Hill, one-half mile above Centerville, a granitic sand, which may possibly also belong to this lacustrine series, forms the bed rock.

Volcanic rocks.—A small dike of andesite occurs a short distance west of Grimes Pass, and fragments of obsidian have been found in Muddy Creek.

THE VALLEY OF GRANITE CREEK.

Configuration.—Granite Creek, joining Grimes Creek 3 miles below Centerville, extends from this point in a north-northwesterly direction for a distance of 8 miles, heading at the low pass north of Quartzburg. A number of tributaries join it from east and west, producing, in the middle course of the creek, a large, basin-like depression. North of Granite a narrow canyon begins. Ophir Creek, Boyles Gulch, and Wolf Creek head at the relatively low divide toward the Payette drainage, while Fall Creek, Trail Creek, and Canyon Creek head in the Boise Ridge and carry a great deal of water. A broad ridge, only 400 feet high, separates Granite Creek and Grimes Creek.

Present stream gravels.—As in the districts already described, the larger part of the present stream gravels consists of tailings. They are one-fourth mile wide at the junction with Grimes Creek, but only a few hundred feet in width from that place to half a mile below the mouth of Ophir Creek. Here they widen and attain a maximum width of 1,500 feet. The tailings also reach far up on Ophir, Boyles, and Wolf creeks, but narrow down to 100 feet or less a short distance above Granite. Along the lower courses of Canyon Creek, Trail Creek, and Clear Creek, as far as 2 or 3 miles above their mouths, are stream gravels and low terraces, attaining in some places a width of 1,000 feet. These creeks have yielded scarcely any gold, and their gravels and terraces show well the character of the river courses of the basin before the gold discoveries. Clear Creek was entirely barren in its upper course. A little gold was found in one or two places along the creek which Hawkins toll road follows from Star ranch westward. Canyon Creek has produced a little gold, and Fall Creek a considerable amount. The main creek has been rich from the junction up to Quartzburg, while no gold is found above that town. Ophir, Boyle, and Wolf creeks were very rich up to their headwaters. The east fork of Alder Creek, which flows into the Payette and heads opposite Wolf Creek, has also produced some placer gold. It appears to be the only creek on the Payette side which is auriferous. The gulches running into Ophir Creek from the east have produced scarcely anything. In fact, most of the broad ridge separating Grimes and Ophir creeks is singularly barren. As to the working of the tailings in Granite Creek, the same remarks apply as have been made in the case of the other tailings mentioned above.

Bench gravels.—Benches are found at intervals all along Granite Creek as far up as half a mile above Granite; also for some distance up Ophir Creek, and up Wolf Creek as far as Placerville. There are usually two well-pronounced horizons, the bed rock of the lower one being 20 to 30 feet above the creeks and that of the higher one at an elevation of 50 feet. Near Granite and Placerville bowlders of porphyritic rocks are abundant. The benches opposite Granite also contain many basaltic bowlders and pebbles, which are probably derived from the basaltic areas high up on the summits of the Boise Ridge, referred to later. On the broad flood plain of Fall Creek the bench gravels are very extensive, and reach an elevation of 4,380 feet above the sea, or 50 feet above Fall Creek. All of the benches along Granite Creek and its tributaries from the northeast have been worked for gold.

Older gravels.—Near Placerville and Granite are several very interesting occurrences of gravels belonging to a stream system which differed considerably from the present one and giving evidence of having undergone disturbances since their deposition. Two small and isolated gravel patches occur close together one-fourth mile northwest of Placerville, on Sailor Gulch. Each contains but a few acres. Both have been worked by the hydraulic process. The exposures show a 30-foot bank of medium-sized gravel with excellent fluviatile stratification. The gravel contains a great number of porphyry bowlders similar to the rock cropping near Quartzburg. The granitic bed rock slopes gently westward, and extends down to the present creek. The bed rock is 150 feet above the level of Wolf Creek at Placerville. A small area of similar gravel lies at the same elevation near the Pioneerville road, 1 mile east-northeast of Placerville.

Opposite Placerville lies the Ranch Company's claim, which has been extensively washed of late years and has produced much gold. This is a mass of older, compact gravels resting in a channel-like depression, and on which, along the creek, the later bench gravels have accumulated. In the early years this gravel was considered "false bed rock," and it was not generally supposed that it could be profitably worked. The gravel body lies on the ridge separating Boyles Gulch from Wolf Creek, the top of it being one-fourth mile wide and reaching an elevation of 200 feet above the level of the tailings at Placerville. It fills a depression or channel, the bed rock rising rapidly northward and southward. The lowest bed rock is exposed in the present diggings, and is at the bank 75 feet above the tailings at Placerville, sloping from there gradually down to the débris-filled stream-bed at the rate of 6 feet per 100 feet. The character of the gravel is shown in fig. 61. The gravels bear every evidence of having been accumulated in a stream of considerable size. They are coarse at the bottom, very well rounded, and contain abundant cobbles of the peculiar altered porphyry occurring near Quartzburg. No basalt bowlders were found. The gold is fine and evenly distributed through the mass of the lower gravel. On the eastern side of the ridge there are also hydraulic pits, and the deepest bed rock lies 135 feet above the creek bed at Placerville. At this point the channel suddenly

18 GEOL, PT 3——43

ceases. Across Ophir Creek the bed rock rises rapidly in all directions, and no possible continuation can be suggested except down the
narrow canyon of Ophir Creek. Furthermore, the nature of the pebbles indicates that the stream came from some point to the northwest,
whereas the present grade of the bed rock is in exactly the opposite
direction. The conclusion is almost unavoidable that the channel
has been cut off by a fault and its grade reversed.

On the point between Wolf Creek and Boyles Gulch a little of the
same compact gravel occurs just at the edge of the tailings. Placerville appears to be built on a similar compact gravel, in which narrow
benches were cut by the present stream, though just below the town
granite bed rock appears at the edge of the tailings. Due southwest
of Placerville, on the slope up to the summit of the level ridge between
Granite and Wolf Creek, no bed rock appears to have been found.
Neither has there been any bed rock found on the opposite slope from
this ridge down to Granite Creek. A broad, low bench fringes the
northeast side of Granite Creek below Granite. Bed rock has never
been found for some distance below Granite in the creek. All this
appears to indicate
that the Ranch
Company's channel
continues with grad-
ually sinking bed
rock across Granite
Creek.

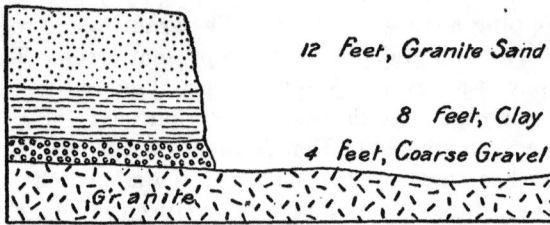

12 Feet, Granite Sand

8 Feet, Clay

4 feet, Coarse Gravel

Granite

FIG. 61.—Bank at the Ranch Company's claim, Placerville.

Interesting con-
ditions also obtain
across Granite Creek
on the broad flood plain of Fall Creek, where it emerges from the narrow canyon in the Boise Ridge, here suddenly rising as a steep escarpment facing west. This place, called Norwegian Flat, at present
worked by the Kennedy Company, is extensively covered with rich
bench gravels up to the mouth of the canyon. This bench gravel, which
is rarely over 25 feet in thickness, does not rest on granite, but on a
harder, more compact gravel, which contains some gold and which is
of the same general character as the Ranch Company's gravel. At
the upper end of the claim, near the mouth of the canyon and the foot
of the escarpment, this "false bed rock" is found to suddenly abut
against the granite, strongly suggesting that it is cut off by a fault.
No shafts have ever been sunk in this lower gravel. It is probable
that it is the same channel continuing across Granite Creek, and that
it is cut off at both ends by faulting. On the hills to the right and
left of Fall Creek a similar compact gravel also occurs, seemingly continuous with that of the deepest channel. Thus the low ridge between
Fall and Canyon creeks is covered, up to an elevation of 80 feet above
Fall Creek at the upper end of Kennedy's claim, and a similar body

reaches high up on the ridge between Fall and Granite creeks nearly to the Newburg or Lawyer quartz claim. It is said to be quite rich here, though hard, and difficult to work on that account and because of the scarcity of water. On this ridge the older gravel distinctly rests on the granite, and its thickness reaches a maximum of 200 feet. Its highest elevation on the ridge near this quartz mine is 4,900 feet, or 500 feet above the upper end of Kennedy's claim.

Gravel on the Boise Ridge.—A very remarkable occurrence of gravel was found on the summit of Boise Ridge 3 miles west-north-west of Quartzburg. The broad, flat summit extends northward from Hawkins Pass, with a winding direction, caused by the deeply incised gulches, and an elevation of from 6,500 to 7,300 feet above the sea. North-northwest of Quartzburg lie, on the summit, several hundred feet of scoriaceous basaltic flows, which extend down into Jerusalem Valley on the west. In this basalt there occurs, at an elevation of 6,900 feet, one-half mile south of the sharp point, with an elevation of 7,200 feet, a small body of gravel. It could not be determined whether this gravel is intercalated between the basalt beds or forms an inclusion in the same, but the former is more probable. At any rate, the gravel is well washed, of granitic character, and is said to contain some gold. This occurrence is of the greatest interest, as it clearly indicates that a very great disturbance has taken place since the eruption of these basalts, for the gravel could not possibly have been formed with the present topographic features. Considered in connection with the faulted old gravels at the eastern foot of the escarpment, 2,700 feet lower, it points to a great disturbance along the eastern base of the Boise Ridge.

Basalt.—The whole of the high portion of the Boise Ridge south of the latitude of Quartzburg is remarkably free from any late eruptions. At the locality mentioned in the preceding paragraph, culminating in the hill with an elevation of 7,200 feet north-northwest of Quartzburg, the very even, flat surface of the granitic ridge is capped by 300 feet of massive basalt. Above the massive basalt lie 200 feet of tuff and scoriaceous basalt, again covered by 100 feet of fresh and massive olivine-basalt. These basaltic flows slope rapidly westward, and the ridge on the north side of Brainard Creek is, as seen in Pl. LXXXIX, composed of a great number of superimposed thin basaltic flows, all dipping northwest at angles of 30° or 35°. The photograph is taken from a gap in the ridge between Brainard and Porter creeks, 3½ miles west-southwest from the high basalt peak in the Boise Ridge. Taken in connection with the probable fault along the eastern side of the range at Granite, it certainly appears as if a westward tilting had formed part of the orographic movements in this vicinity since the time of the deposition of the lake beds. The basalt is much older than the basalt of the lower Moore Creek, and in all probability is contemporaneous with the lake beds.

Other small basalt areas were noted near Hawkins toll road, 2 miles west of Star Ranch. This basalt forms a dike on the high ridge between the toll road and Clear Creek, follows the road for some distance near longitude 116°, and appears again on the ridge north of the road. Here it is in part vesicular, and appears to have flowed out from the dike. This basalt is probably also of Tertiary age.

FINENESS OF THE GOLD.

The fineness of the placer gold varies from 770 to 912. The following list gives the value at some of the principal localities:

Value and fineness of placer gold at principal localities.

Locality.	Value before melting.	Fineness per mill.
Idaho City, Turner's claim (bench gravels)	$16.75	
Idaho City, East Hill (high gravels)	16.50 to 16.75	
Idaho City, Barker's claim (high gravels)		850
Pioneerville (bench gravels)	15.35	770
Placerville, Ranch Company claim (old channel gravel)		912
Placerville, Granite Creek (bench gravel)		850
Placerville, Fall Creek		775
Placerville, Ophir Creek		910
Placerville, Wolf Creek		910

WATER SUPPLY.

The water available for placer work is not abundant; in fact, usually the working season is only from three to four months. The streams are not large and do not head among very high mountains, so that the flood from the melting snows soon subsides. It would be possible, but hardly practicable, to carry a ditch to the basin from the headwaters of the Boise or the Payette. The principal ditches are as follows:

MOORE CREEK.

1. The upper Moore Creek ditch. Takes water from Moore Creek 5 miles above Gambrinus Gulch. Capacity, 600 miner's inches.
2. The Lambing ditch. Dam below Illinois Gulch. Capacity, 350 miner's inches.
3. The Christie ditch. Dam above mouth of Granite Creek. Capacity 1,100 miner's inches. These three are on the northwest side of Moore Creek.
4. Thorn Creek ditch. Takes its water from Thorn Creek, dumping it across the divide into Pine Creek; used for the high gravels south of Idaho City.
5. Channel ditch. Dam on Elk Creek one-half mile above Elkhorn Mill. Capacity, 600 miner's inches. Covers divide between Elk and Grimes creeks.
6. Mann's ditch. On south side of Elk Creek. Dam 1 mile above Forest King Gulch. Ten miles long. Covers Gold Hill at Idaho City.

7. Anderson ditch.   Takes water from Elk Creek below Boulder mine.
8. Cuddy ditch.   Takes water from Elk Creek below Boulder mine.
9. Dunn ditch.   Takes water from Elk Creek below Boulder mine.

GRIMES CREEK.

The water supply is here largely controlled by the Wilson Company. The principal ditches are as follows:

1. Wilson ditch.   12 miles long, on west side of Grimes Creek.   Dam located 1 mile below Charlotte Gulch.   Capacity, 2,000 miner's inches.
2. Gold Trap ditch.   7 miles long, also on west side of Grimes Creek.
3. American ditch.   6 miles long, west side of Grimes Creek.   Dam at Charlotte Gulch.
4. Mountain ditch.   8 miles long, west side of Grimes Creek.   Also two shorter ditches from Clear Creek.

GRANITE CREEK.

The water supply is here controlled by the Ranch Company and the Kennedy Company.   The Ranch Company's ditch is 14 miles long, and takes water from Granite Creek above Quartzburg.   Capacity, 1,000 miner's inches.   Other ditches lead from West Fork, Fall Creek, and Canyon Creek.

The hydraulic washings use a pressure of from 100 to 350 feet. There are rarely more than 600 miner's inches used in each monitor. Hydraulic elevators have been used by the Wilson Company to work low-bench gravels and tailings.   The gravel is elevated from 10 to 25 feet.   Five hundred miner's inches are here used for the monitors.

GROUND AVAILABLE FOR FUTURE WORK.

The largest amounts of gravel remaining near Idaho City are on East Hill and Gold Hill.   Though some of the bench gravels remain they are getting rapidly worked out.   In the Grimes Creek drainage some bench gravels still remain near Pioneerville.   In the Granite Creek drainage there is a considerable area of low-bench gravels yet untouched near Granite, on both sides of the creek.   Considerable gravel remains at the Ranch Company's ground near Placerville.   If the conclusions in this paper are correct, there is also a large amount of gravel on the ridge between Wolf Creek and Graniteville.

In all three drainage branches there are vast amounts of tailings which, with suitable appliances, may be worked over again.   But there is no doubt that the placers in course of time will be exhausted. That they are not already exhausted is due to the limited water supply

THE MONAZITE SANDS.

The sand of the gravels and lake beds of the Idaho Basin is entirely derived from the granite and associated dike rocks.   It consists of relatively angular and sharp-edged grains, indicating its manner of formation by extremely rapid accumulation from the deeply disintegrated rocks. The heavy minerals found in the granite are also found in the sand, and may be easily separated by washing in the miner's pan; they are always deposited in the sluice boxes with the gold.

In all parts of the basin a yellow or brownish-yellow mineral forms a considerable quantity of the heavy substances remaining with the gold. It is usually referred to as "yellow sand," and is also given the picturesque name of "Bummer Hill sand," from a locality near Centerville, where it was particularly abundant, but I am not aware that its true character has ever been investigated.

The mineral has been shown to be monazite, this being the first time its occurrence has been noted from the Western States. As is well known, it occurs abundantly in the granite and gneissoid rocks and gold-placer mines of the Southern Appalachians, and in several of the North Atlantic States, also in Brazil, the Ural Mountains, and other places. There is no doubt that it forms an original constituent of the granite of the Idaho Basin.

One of the samples was obtained in washing a few pans of the sandy lake beds occurring as "false bed rock" in a gravel bench at the junction of Moore Creek and Granite Creek, 3 miles east of Idaho City (see fig. 59). The heavy residue consisted largely of small yellow grains and amounted to about 2 grams per pan of 8 kilograms, which would correspond to 0.025 per cent. The microscope revealed the following minerals: Ilmenite in sharp hexagonal crystals, but no magnetite; zircon, also in extremely sharp crystals of a slightly brownish color, and abundant yellow or greenish-yellow grains rarely showing crystallographic faces. The refraction and double refraction of the latter mineral were very high; the hardness not much over 5. The ilmenite was eliminated by the electro-magnet, and the remaining powder, containing about 70 per cent of the yellow mineral, was analyzed by Dr. W. F. Hillebrand. The result showed it to be a phosphate of the cerium metals, the approximate amount of the oxides of the latter being 48 per cent; in these approximately 1.20 per cent of thoria was found. This result identifies the mineral with monazite, the only other similar mineral being xenotime, which is mainly a phosphate of yttrium with but little cerium. The samples also contained a considerable amount of titanium, which would indicate that some titanite is present. Practically all of the ilmenite was extracted by the magnet.

Another sample, furnished me by Mr. T. Myer, of Placerville, came from the alluvial gold washings in Wolf Creek, near that town. Cleaned from quartz, etc., it appeared as a heavy dark sand consisting of a black iron ore (ilmenite), rounded crystals of red garnet, sharp crystals of zircon, and irregular grains of a dark yellowish-brown mineral with waxy luster, sometimes showing crystallographic faces. It was found impossible to extract more than a small part of the iron ore by the magnet. There was practically no magnetite present. This sand was examined qualitatively by Dr. Hillebrand, who found phosphoric acid, cerium metals, and thorium. The yellowish-brown material is therefore, in all probability, monazite.

Monazite has, as is well known, a certain economic value, as the oxides of the rare earths contained in it are used for the preparation of the incandescent gaslights of the Welsbach and other burners. A considerable amount of monazite sand has been produced during the last few years in the Southern Atlantic States, chiefly North Carolina, and in Brazil. The prices have varied from 3 to 25 cents per pound, according to the purity and the percentage of thoria. The North Carolina monazite contains between 0.17 and 6.20 per cent of this rare earth. Mr. Waldron Shapleigh, chemist for the Welsbach Light Company, kindly gives (March, 1897) the following information in regard to present production and price:

At present the monazite sand market is very dull; hardly any demand, and only in lots of a few tons. The present price in North Carolina is 6 cents per pound, but it can be bought a little lower than this. Brazilian sand is quoted in New York at 4¾ cents per pound, fully equal to the North Carolina sand, in 5 or 10 ton lots. Larger orders can be placed in Brazil at a much lower figure. The price has been steadily downward, as the supply from the mines now opened is far greater than the consumption. It is not generally known that during the first eighteen months or two years of this new industry enough sand was mined and purchased by the largest manufacturers to last several years to come, as so far it has but the one use. The manufacturers did not know how extensive the sand deposit was; therefore were desirous of securing a large and safe stock at the start.

The largest purchasers are the Welsbach Light Company of Vienna, supplying Europe, and the Welsbach Light Company of Philadelphia, which supplies America. I should hardly think that Western localities could compete with North Carolina and Brazil, unless the mineral is of a very superior quality or a by-product.

At present there would be no difficulty in placing an order for several thousand tons per year in Brazil and having it filled.

The widespread occurrence of monazite in considerable quantities in the Idaho Basin raises the question whether the deposits can be profitably worked. The present low price and the high cost of transportation and labor make this very doubtful, unless it be saved as a by-product in the placer mines. It will be necessary to extract all ilmenite by strong electro-magnets. In this manner a comparatively pure product may be obtained. It is not practicable to entirely separate the zircon and garnet from the monazite. The purest material was obtained from near Idaho City, while that from Placerville and vicinity contains a large amount of ilmenite and garnets. Many data in regard to the character and production of monazite may be found in a paper by Mr. H. B. C. Nitze, in the Sixteenth Annual Report of the United States Geological Survey (1894–95), Part IV, pp. 667–693.

RELATION BETWEEN PLACERS AND QUARTZ VEINS.

The dependence of the gold placers upon the occurrence of gold-quartz veins is very strongly brought out by a study of the occurrences of both. While there are many small quartz seams occurring throughout the granite, some of which may contain a little gold, it is perfectly evident that there are two regions in which quartz-vein deposits are concentrated. These are, first, the Gambrinus mining district, on the ridge between Elk Creek and Moore Creek, continued by the Elkhorn mining district, at the headwaters of Elk Creek; and second, the gold belt extending from the Boise Ridge near Quartzburg to Grimes Pass. Every creek and ravine leading up to these deposits has been rich, while the watercourses rising in other parts of the range are comparatively barren. It seems clear beyond doubt that most of the gold in the gravels near Idaho City came down the Illinois and Gambrinus gulches; above these Moore Creek becomes comparatively poor. What gold there is has doubtless been derived from the more distant veins at the very head of Moore Creek, near Summit Flat. In nearly every case an exceptionally rich ravine has been found to lead up to a quartz vein. Thus it is clear that the recently discovered Summit mine, on the ridge between Elk and Grimes creeks, furnished the gold found in Deer Creek and Henry Creek. The only occurrence to which some doubt is attached as to the derivation of the gold is that of the rich angular gravels at the head of Spanish Fork and Willow Creek, but it is probable that they are derived from local seams and veins.

The headwaters of Grimes Creek furnish a most convincing argument in favor of the derivation of the gold from the quartz belt. Every gulch heading along the line of that belt is rich, while every one not crossing it carries only extremely small quantities of gold.

The conclusion is that practically all of the placer gold in the district has been derived from the quartz veins in these two districts.

On an average the fineness of the gold in the quartz veins is a little less than that in the placers. This has generally been the experience in most mining districts, and is accounted for by a dissolving of the silver and baser metals from the surface of each grain of gold. The highest grade of quartz gold is found near Quartzburg, and in the gravels of Placerville, derived from the Quartzburg mines, the placer gold is of unusual fineness.

CHAPTER IV.

THE IDAHO BASIN (CONTINUED).

THE PRE-TERTIARY ROCKS.

GRANITE.

The Idaho Basin forms a part of the great granite area of the Boise and Payette drainage, and the pre-Tertiary rocks consist exclusively of granite, together with a number of porphyritic dikes, probably intruded shortly after the granitic intrusion. It has been stated before that the age of this granite is unknown. If Archean, as has been supposed, the slight amount of compression and change it has undergone is certainly remarkable. A shearing is often noticeable, dividing the rock into sheets or plates upward of a foot thick. The direction of the shearing varies considerably, and is sometimes parallel to the general direction of the quartz vein. In Gambrinus and Sub-Rosa gulches a strike of N. 20° to 24° W. and a dip of 70° E. or W. were noted. In the Ranch Company's claim at Placerville the sheeting is parallel to the direction of the Quartzburg gold belt; strike, N. 45° E.; dip, 60° SE. On Hawkins toll road, on the western slope of the Boise Ridge, the strike is N. 60° to 70° W. and the dip 70° to 80° N. or S. Conjugated systems of shear planes, having the same strike, but dipping in opposite direction, thus occur here. In the depressions and low ridges of the basin the granite is disintegrated to considerable depth, so that it is very difficult to secure good specimens. The disintegrated granite forms a coarse, yellowish-gray sand, the individual grains of which have undergone but very little decomposition, and which is easily swept down into the creeks by the rain storms. Fresher and harder rocks crop out on the high ridge between Elk and Moore creeks, upon which the Forest King and other mines are located. The outcrops form brilliant white rounded masses; but even here the disintegration has made rapid progress. On the Boise Ridge the granite is ordinarily soft and crumbling. The deepest disintegration is probably found about the head of Muddy and Ophir creeks, where a good outcrop is only rarely seen. The granite area extends northward across the Payette River and far to the north of it.

The granite has a coarse grain, the average size of the constituents being 3^{mm}. The reddish orthoclase crystals are often very prominent.

The rock is composed of gray quartz, white or reddish feldspar—partly orthoclase, partly an acid soda-lime-feldspar—and small biotite flakes, the incipient decomposition of which usually gives the rock a rusty aspect. Hornblende is rarely found. Muscovite often occurs in the more acid varieties. Of accessory constituents, which are best studied by washing the decomposed sandy granite, there are: Ilmenite in often perfect crystals (though little or no magnetite appears to be present), apatite, zircon in extremely sharp, slightly brownish crystals, small garnets, titanite, and brownish or yellowish imperfect crystals of monazite. For description of the monazite sands, see pp. 677–679.

DIKES ASSOCIATED WITH THE GRANITE.

The granite is traversed by dikes, which in some places become very numerous and large. Dikes of granite-porphyry and aplite are common, though rarely very long and wide. Many such dikes, together with others of pegmatitic character, occur; for instance, along the road southwest of Idaho City and on the hill, with an elevation of 6,200 feet, due south of the town.

Dark-colored, lamprophyric dike rocks, which generally belong to the minettes, were noted in a few localities. These dikes are, as a rule, narrow, and their occurrence is closely connected with that of the veins. From the Sub-Rosa and Forest King mines dark-gray, fine-granular dike rocks were collected, generally rich in black mica. In thin section the former appears as typical minette, consisting of biotite, augite, magnetite, and orthoclase, with panidiomorphic structure. The feldspar crystals show a tendency to radial or spherulitic arrangement. A similar minette, consisting of biotite, hornblende, and orthoclase, was collected at the Gold Dollar tunnel, 1,000 feet east of the pass leading from Placerville to Garden Valley. The dikes evidently antedate the veins.

The Boise Ridge south of Quartzburg contains scattered dikes of granite-porphyry and diorite-porphyrite, but near the latter locality begins a very important belt of dikes intimately connected with the Quartzburg belt of gold deposits. The rocks are in all respects similar to those which appear at Willow Creek mining district; the latter may in fact be regarded as the westward extension of the Quartzburg belt, having the same direction and lying in its continuation to the west-southwest, but a distance of 8 miles, barren of mineral deposits and dikes, separates them. The dikes do not follow the mineral deposits in detail and are very irregular, sometimes only a few hundred feet wide, then again expanding to a width of over a mile. Owing to unusually deep residuary soil on the divide toward Payette River, the contact of granite and porphyries is generally difficult to trace, and the areal extent indicated on Pl. XCVI must be regarded as only approximately correct.

The porphyry begins at the Mountain Chief and Belzazzar mines, where it occurs as a wide belt extending across the vein and forming a considerable part of the hill to the east of the mines. Large masses of a similar porphyry occur in the lower part of Fall Creek, and it continues eastward toward Quartzburg as a narrowing belt. The rock in this area is a characteristic light-colored hornblende-porphyrite, consisting of white plagioclase in stout prisms up to 1^{cm} long, and idiomorphic hornblende crystals up to 5^{mm} in length, embedded in a fine-grained groundmass of feldspar, hornblende, and a little quartz.

At Quartzburg the porphyry belt is only a few hundred feet wide and has undergone great thermal alteration, secondary minerals like pyrite, muscovite, and calcite being abundant. On the Gold Hill mine dump specimens of fairly fresh rocks were collected. Some of them are very similar to the above-described rock. The abundant large white feldspars are labradorite, according to extinction of numerous Carlsbad twins. There is no hornblende left undecomposed, but there is some light-brown biotite in process of conversion to chlorite. The groundmass is micropoikilitic, consisting of quartz and unstriated feldspar. Another variety, greatly altered, of a yellowish color and impregnated with pyrite, is characterized by large corroded quartz crystals, up to 1^{cm} in diameter, and large porphyritic labradorite, greatly altered by sericitization. The groundmass is fine-granular, probably micropoikilitic, of quartz and feldspar, but now greatly filled with sericite.

East of Gold Hill the porphyry belt widens considerably. Quartzburg Hill is composed of quartz-diorite-porphyrite. This rock, similar to that described from the Belzazzar mine, forms the largest part of the area. It crosses Wolf Creek as a wide belt, and is here accompanied by dikes of more basic rocks. Two miles north of Placerville a dike of norite-gabbro crosses the road and a dike of normal diabase appears in the same vicinity. East of Wolf Creek the area widens still more and reaches its maximum width. At Sweet's Claim, one-half mile west of Grimes Pass, the porphyry is narrower. A specimen from the Northern Star shaft shows the same idiomorphic large feldspar and hornblende crystals. The former are very fresh and consist of labradorite; the latter are sharply idiomorphic and partly chloritized. The groundmass is microcrystalline allotriomorphic, being made up of quartz and unstriated, clouded feldspar.

This porphyry belt was not followed east of Grimes Pass, but it is apparent that it attains great development among the high hills rising beyond it. Clear Creek contains a great abundance of porphyry bowlders. They are practically the same quartz-diorite-porphyrite which is described above. Quartz is evidently always present in the groundmass, and sometimes also as porphyritic crystals.

THE QUARTZ VEINS.

THE IDAHO CITY GOLD BELT.

In the immediate vicinity of Idaho City very few quartz veins occur, and none of importance. Mr. Plowman states that a big quartz vein was found on Wallula Flat, 2 miles east of Idaho City, on the northern side of the creek. A narrow streak in this vein carries gold. About 1 mile due south of Idaho City the Keystone mine was located on a large but apparently barren vein.

Six miles northeast of Idaho City, on the ridge between Moore Creek and Elk Creek, is the Gambrinus mining district.

The Blaine vein is situated on the Moore Creek side of the ridge. The developments are small, though some fair ore has been extracted and milled. There is a 5-stamp mill on the property.

The Chickahominy vein lies a little higher up on the ridge and a few hundred feet south of the Blaine. It was worked in the early days, and a large mill was erected on Moore Creek 4½ miles above Idaho City.

The Illinois vein.—This important deposit, which can be traced for 1¾ miles, is located three-fourths of a mile above the Blaine, and crosses the ridge at an elevation of 5,100 feet. The vein has an average strike of a few degrees north of west and dips to the south at an angle of from 45° to 50°. It is inclosed in granite throughout, though smaller dikes of porphyry occur occasionally in this granite. The vein is one of the strongest in the basin, and it is very clear that it has furnished a large portion of the placer gold in Moore Creek. It contains a great deal of finely distributed gold all along, and some good pay shoots besides. Illinois Gulch, draining this vicinity, is reported to have been extraordinarily rich.

The Eureka claim lies at the eastern end of the vein. Its production amounts to $30,000, and besides much was taken from surface diggings near the vein. A 10-stamp mill stands on the claim, which has not been worked since about 1880. The vein is a large, composite one, similar to the Illinois, to be described later. At the time of its exploitation an ore shoot in this vein was worked down to a small depth, where it is said to have been lost. The developments are slight, consisting only of a tunnel and a shaft 60 feet deep.

The Lucky Boy adjoins on the west. The developments consist of only a few prospect holes. The strike is N. 81° W. and the dip 50° S. The exposures show 8 feet of decomposed and sheeted granite, with small quartz seams. Some pay ore is found on the foot wall. Two parallel veins exist, one 30 feet south and the other 200 feet north of the main fissure.

The Illinois, consisting of two claims, is the principal producer of the vein. It was located in the early days, as it was soon seen that

the rich placers led up to the vein. The production is stated to be
$225,000, which has been chiefly taken out in small batches of very
rich ore. Some ore from this claim was crushed in 1895 in the Blaine
mill. The deposit is characteristic of many veins in the Idaho Basin.
Large masses of quartz are hardly ever seen. The strike is N. 71° W.
and the dip 40° S. The vein consists of a wide zone of sheeted and
fractured granite, with abundant small quartz seams between the
joints of the sheets or ramifying through them. These small quartz
seams carry the gold, while the granite between them, though usually
altered, soft, and decomposed, contains no pay. This sheeted zone
is here 30 to 40 feet wide and contains gold throughout. The principal pay shoot is, however, 400 feet long, and one streak, 2 feet wide,
in this pay shoot is particularly rich. The deposit has been sluiced
off on the surface and then worked by means of a crosscut tunnel
from the gulch. A cross seam, also carrying some gold, joins it in
Illinois Gulch, and a parallel vein is also said to exist. The ore is
practically all free milling on the surface. Even when sulphurets
are met with in depth it is probable that a larger proportion of the
gold will remain in free condition. The developments may be said to
be very slight and unsatisfactory, but it is probable that if properly
opened the vein would furnish great amounts of low-grade ore, which,
with suitable and cheap methods of extraction and reduction, might
be made to pay. The width of the vein and the soft character of the
rock would make mining somewhat expensive and difficult.

The Chicago claim, on which but a slight amount of development
work has been done, adjoins on the west, beginning near the road on
the summit of the ridge.

The Populist vein.—This vein is situated 1 mile west-northwest of
the Illinois. It has a similar strike and dip, and some work has been
done on it recently.

The Cleveland vein.—This relatively small vein is located 5 miles
north-northwest of Idaho City, on the south side of Forest King Gulch,
at an elevation of 5,030 feet. It is a single-fissure vein, about 1 to 2
feet wide, inclosed in granite, striking a little north of west and dipping 60° S. A rich shoot 100 feet long was found on it a few years
ago, which yielded a considerable sum of money; it has not been
followed below the present tunnel level. The mine is of interest as
showing plainly the result of faulting, illustrated in fig. 62. The vein
is thrown a distance of 60 feet horizontally in the hanging wall. At
least two faults of this character are known. If movements in only
a vertical direction have taken place, this fault is certainly an overthrust, the hanging wall having moved up relatively. As, however,
lateral movements may also have occurred, it is possible that the
present position of the vein may be due to an oblique movement to
the southwest, the direction of which makes with the horizontal an
angle of between 0° and 60°, the hanging wall having moved down

relatively. While this is possible, the probabilities are strongly in favor of an overthrust.

The Gambrinus (Surprise) vein.—This large deposit is situated 5½ miles north-northeast of Idaho City, on the summit of the ridge between Elk and Moore creeks and near the foot of the steep Forest King Hill. The elevation is 5,480 feet. It was discovered in 1864, and the principal work was done between 1864 and 1865. The total production is $263,000. The strike of the vein is N. 61° W., and the dip, measured on the exposed foot wall, is 45° S. It can be traced for some distance westward, and the Buckeye forms the western extension. The vein has a maximum width of 40 feet, and is of the same composite type as the Illinois, consisting of a sheeted zone in granite with a great many small and rich quartz seams between the joints, these seams containing all the pay. The surface is very greatly decomposed and has been extensively sluiced. Two chimneys or shoots of rich ore were found close together, dipping east on the plane of the vein, at first steep and then more gentle. These shoots were not nearly so long as the Illinois ore body, but richer. The ore contains antimonite, and this mineral appears to be closely associated with the gold. The vein has been developed only by short tunnels and small shafts, not over 100 feet deep, and appears, like the Illinois, to deserve much more extensive prospecting.

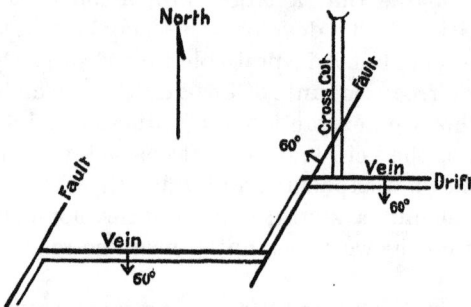

FIG. 62.—Diagram of fault in the Cleveland vein.

The same difficulties in the way of water and heavy ground will probably be met with in depth.

The Boulder vein.—This deposit is located on Elk Creek, 6½ miles north by east of Idaho City, at an elevation of 4,830 feet. Preparations were made a few years ago to exploit this vein on a large scale, and an excellent 30-stamp mill, to be driven by water power, was built. It was run for only a short time, the ore probably being of too low grade to handle profitably; neither was there much ore in sight, so that it would have been necessary to sink below the creek level almost immediately. The vein, which crops for only a short distance and is marked on the canyon slope by quartz croppings several feet thick, strikes on the surface N. 75° W. and dips 50° S. It is opened by a tunnel 900 feet long. The general character is the same as the Gambrinus and the Illinois, being a very wide (up to 40 or even 60 feet) zone of sheeted granite filled with many small seams of quartz carrying pyrite, arsenopyrite, and zincblende scattered through it. As it was necessary to mine the whole width, though the pay was only in the narrow seams, and as the ground was very soft and heavy, square timbering had to be used. Underground the vein appears to strike N. 20° W., from which disagreement with the surface strike it would appear that the vein has been considerably disturbed. A narrow vein in the hanging wall of the big deposit was mined during 1896 and furnished small quantities of very rich ore. This was not much decomposed, and indicated what the character of the ore will be in depth. It contained bunches of pyrite, arsenopyrite, and blende, with a gangue of quartz and some calcite. The fresh sulphurets contained much coarse gold, the ore being free milling to 70 per cent of its value. It is probable that most of the vein in the Gambrinus mining district will remain largely free milling as depth is attained. The value of the coarse gold in the sulphurets is $15 to $16. The fineness of the amalgamated bullion is 680 to 718.

Mona MacCarthy is the name of a small vein located three-fourths of a mile west of the Boulder, on the high side, and lying between the Boulder and the Forest King veins. Some work has been done on it. Many small seams are found between this vein and the Forest King.

The Sub-Rosa or Forest King vein.—This vein is traceable for a distance of 2 miles from the Washington mine on the east to beyond the Forest King on the west. It crops in hard granite throughout, though at several places dark-green, dioritic dikes appear near it or cross it. Though narrower than the Illinois and the Gambrinus, it has produced some good pay shoots.

The Forest King was located in 1875, and a 10-stamp mill was erected on it in 1884. The elevation of the mill is 6,280 feet. The United States mineral monument of the district is indicated by an iron rod in an outcrop close by. No work has been done on the vein during the last few years. The vein shows on the surface in quartzose croppings, which do not contain any pay. It is opened by a tunnel 900 feet long, through granite, which shows several seams dipping southward. The vein consists of a zone, several feet wide, of crushed granite with smaller quartz seams. It strikes N. 56° W., and dips 60° S. The drift on the vein extends several hundred feet east and west. Three hundred feet west a dike, 20 feet wide, of a dark, syenitic rock (probably a minette) apparently cuts across the vein; but from the fact that the dike is full of small quartz seams, which all contain a little gold, it is probable that the dike is really older than the vein, and that the difference in appearance of the vein in the two rocks is due to the difference in their resistance to the dislocating force. A dike of the same rock, 1 foot wide, occurs in the crosscut, and is parallel to the vein in dip and strike. A 50-foot winze was sunk below the tunnel level, some distance west of the crosscut, in the bottom of which was found altered granite, with sulphurets and small quartz seams, giving assay values of $80 to $100 per ton.

A location called the Northern Light, on which some work has been done, adjoins the Forest King on the east. The Sub-Rosa, also called the Confederate, is a claim on the same vein, 1,500 feet long, and located on the steep side hill toward Moore Creek. The topography is very rugged, owing to the depth to which the gulches have been incised in the granite. At the line between the Sub-Rosa and the Washington the elevation is 5,500 feet. The vein was worked many years ago by Mr. William Hooten, who extracted a considerable amount of very high grade ore from a comparatively small shoot. Then the mine lay idle for many years, until 1896, when work was resumed with the intention of finding the continuation of the pay shoot from a lower tunnel level. The strike of the vein is N. 56° W., and the dip is, as usual, to the south. The vein is several feet thick, consisting of a very soft clayey mass of altered granite with quartz seams. Many smaller dikes of lamprophyric rocks, chiefly minettes, occur near the vein. In one of the older tunnels it is clearly seen how one of these dikes is sharply cut off and faulted by the vein. The exposures in the lower tunnel of 1896, 300 feet below the old workings, are interesting. A small portion of the vein was found in about normal position, but on following it toward the west, in the direction of the ore shoot, it was found to be cut off by a dike of minette 20 to 30 feet wide, across which solid granite again was met. At first glance it would appear as if a later dike had cut across the vein and faulted it, but upon close inspection of the dike it is seen to be extremely crushed and separated from the vein by fault planes, and the probability is that the dike was intruded before the vein was formed, and that a subsequent fault has thrown the vein in the hanging wall, just as happened in the Cleveland vein. This is made the more probable as extensive explorations had previously failed to find it in the foot wall.

The Washington claim adjoins the Sub-Rosa on the east. This part of the vein was exploited a few years ago, and a considerable amount of gold was extracted. The mine is equipped with a 10-stamp mill, and is developed by a tunnel following the vein for 290 feet and a vertical shaft sunk to 316 feet at the mouth of the tunnel; three levels are turned from the shaft, and extend, the first to 400, the second to 250, and the third to 170 feet toward the east. The vein is vertical, and has about the same strike as the Sub-Rosa. An ore shoot 45 feet long and from 1 to 6 feet thick was found, and has been stoped from the 200-foot level up to the surface. The yield is reported to have been $90,000 from 4,300 tons, or $20 per ton. The ore was practically all free-milling and consisted of fresh quartz. A little pyrite occurred in depth. The shoot was cut off in depth by a small vein carrying silver, and its continuation beyond this is not known. Forty feet north of the gold vein, and separated from it by altered granite, is a strong vein of solid quartz, from 8 inches to 4 feet wide, which has been exposed by crosscuts from all levels. This vein carries silver only—as chloride on top, stephanite and ruby silver in depth—the average assays showing values of from 33 to 90 ounces per ton. Ore of the latter kind carries only $1 of gold. Though much of this silver ore is in sight, none has yet been extracted. Four hundred and fifty feet beyond the breast of the tunnel another ore body is said to show on the surface and to carry both gold and silver. This is one of the few occurrences of silver veins in the basin, and is of great interest, as the two veins evidently represent separate periods of vein filling, the silver vein probably being the later.

The Elkhorn mining district adjoins the Gambrinus and is situated on upper Elk Creek.

The Elkhorn vein is an old location at the junction of Elk Creek and Ross Fork, at an elevation of about 5,300 feet. Discovered in early days, it was worked during 1867 and 1868, and, intermittently, later. Some prospecting was done on it in 1896. it has produced a total of $500,000. The developments consist of a tunnel 1,400 feet long, and stopes above it. It is a well-defined vein, about 18 inches wide, with a northwesterly strike, and carryir.g decomposed quartz without sulphurets. The ore body was large and the ore high grade, containing up to $40 per ton in gold. The ore shoot was very large, but at a depth of 100 feet it was cut off by a fault plane carrying soft, decomposed granite. There are several other veins in the vicinity, which, however, can show no production.

The Summit vein is a recent discovery on the ridge between Elk and Grimes creeks, found by tracing the placer gold of Deer Creek up to its source. The vein is inclosed in granite, and strikes a little north of west, dipping 45° SW. There is a zone, 18 feet wide, of crushed granite, carrying 4 feet of pay ore, composed of the same broken granite and quartz seams, and which assays from $10 to $40 per ton in gold. The ore shoot is said to be 60 feet long on the surface. As usual, the vein carries much water and is difficult ground to timber. Prospecting was in progress in 1896, and it was proposed to sink a shaft 400 feet deep. Between the Summit vein and Centerville lie, near the road, two quartz claims, called the Golden Fleece and the Golden Star. A 10-stamp mill was built long ago to work the ore, but the results proved unsatisfactory, and the property has long been idle.

THE QUARTZBURG-GRIMES PASS GOLD BELT.

The whole lower drainage basin of Granite Creek and Grimes Creek is singularly void of gold-quartz veins. Three miles north of Centerville, at Crane's claim, a deposit carrying much sulphurets containing gold and silver is being prospected. On Clear Creek, 3 miles south-southwest of Star Ranch, at an elevation of 4,475 feet, are two narrow quartz veins (Jackson's claim), inclosed in granite and carrying silver only. A small quantity of rich silver sulphides was found

18 GEOL, PT 3——44

on them, and stains of copper and arsenic were noted.   The strike is north to south and the dip steep to the east.

Two and a half miles southwest of Quartzburg the gold belt begins, at the Ebenezer claim.   It is not known to extend west of this point, though gold-quartz float has been found 2 miles farther on, along Dead Man's Gulch.   The direction of the gold belt westward would carry it directly to Horseshoe Bend, where the Willow Creek belt begins.   The high Boise Ridge is very brushy and difficult of exploration, and it is by no means impossible that quartz veins will be found in the intervening stretch.

The Ebenezer vein.—This vein is continuous for a distance of nearly a mile across the gap in the ridge between Canyon and Fall creeks, and three important claims are located on it.   The Ebenezer claim lies on the Canyon Creek side, and is said to have produced $150,000 from sluicing and surface workings.   The vein strikes northeast and southwest and dips to the southeast, and is encased in granite.   It is about 5 feet wide, and consists of sheeted granite traversed by many small and rich quartz veins. The surface ore was very rich, but at a slight depth the gold was contained in sulphides, which did not readily yield it to simple amalgamation. Only assessment work has been done during the last years.

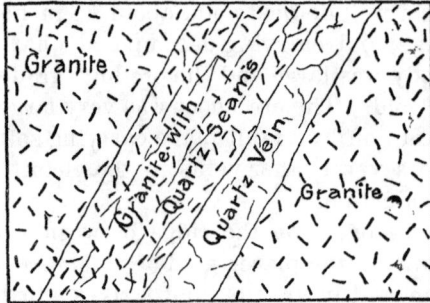

Fig. 63.—Section of Mountain Chief vein, east end of claim.

The Mountain Chief adjoins on the northeast, extending to the summit of the ridge, at an elevation of 6,000 feet, and the vein is similar to the one just described.   It is stated that 10 tons of its ore were milled in 1895, yielding $100 per ton, and much gold has been obtained by sluicing the surface.   Sulphurets appear here also in depth.   In a surface cut the section of the vein was as shown in fig. 63.

The Belzazzar claim lies on the Fall Creek side and has been opened by sluicing and a tunnel, 200 feet below the summit.   Bodies of heavy sulphurets, chiefly pyrite, are exposed along the vein.   The western part of the vein lies in hornblende-porphyrite, while the eastern end has granite in the foot wall and the same porphyrite in the hanging wall.

A slightly divergent vein, called the Centennial, lies a few hundred feet southeast of the Mountain Chief, on the summit of the ridge. This vein carries more silver than gold, and shows heavy iron pyrite in a 6-inch seam.

The Gold Hill vein.—This is probably the continuation of the Ebenezer vein, though it has not been traced across Fall Creek. It is continuous from the Newburg, on the divide between Fall Creek and the west fork of Granite Creek, to at least some distance east of Quartzburg.

The Newburg claim, at an elevation of 5,000 feet, was worked extensively by surface sluicing during the early days, then abandoned, and again located. It is developed to some extent by tunnels aggregating several hundred feet in length. The vein consists of a shattered and decomposed zone in a belt of quartz-porphyrite, and reaches a width of 70 feet. Narrow seams extremely rich in gold traverse this shattered zone, giving to the whole an assay value variously stated from $4 to $12. The ore body is evidently extensive.

The Homeward Bound, Elizabeth, and Mayflower adjoin the Newburg on the northeast across West Fork. Considerable surface work with sluices and arrastres has been done on them. The vein is about 5 feet wide, and is said to contain sulphurets in large quantities in depth. A narrow streak of quartz-porphyrite follows it. The Confederate and the Dunlap adjoin Gold Hill on the west, and have a strike of N. 65° E. The developments are not extensive, but some good ore is reported to occur on them.

The Gold Hill and Pioneer claims constitute the most important quartz mine in the Idaho Basin, and the only one which has been extensively and systematically worked, having been in operation with short interruptions since 1864. The Gold Hill was first worked and yielded for a long time ore averaging $20 per ton. It was first exploited by tunnels on the northeastern side of Granite Creek, but in 1875 work began below water level in a shaft just below Quartzburg on the eastern branch of the creek, the total depth attained being 400 feet. In late years the Gold Hill vein has been abandoned and work concentrated on the Pioneer, a claim adjoining on the southeast. The total amount extracted from the Gold Hill and Dunlap claims from 1869 to 1894 is stated to have been $1,280,000, and the production from the Pioneer claim from 1884 to 1895 is stated to have been $498,000. The total production of the claims mentioned, all of which was not recorded, is believed to have been at least $2,225,000.

In the mint reports the following data are found: Raymond's report for 1872 states that Gold Hill produced $300,000 since September, 1869. In 1881 it is stated that 150 tons of ore from the Gold Hill yielded $25,000 in gold. In 1883 the Gold Hill produced $76,800; in 1884 the production was $50,000. The property was equipped with a 25-stamp mill in 1875. No work was in progress in 1896, pending a sale of the mine. The Gold Hill vein is a well-defined quartz vein with an average strike of N. 70° E., and a dip of 70° S. The foot wall is generally sharp and well defined and consists of granite. In the hanging wall lies a dike of quartz-porphyrite several hundred feet wide, described in connection with dike rocks (p. 682). The vein is of very irregular width, from a few inches up to 6 feet, and has a great tendency to throw out stringers in the porphyry in the hanging wall; these when followed were found to be rich, but soon gave out. The ore consists of free gold and sulphurets, the latter chiefly iron pyrite, also some antimonite, which is said to be intimately associated with gold. Small traces of tellurium are also reported. About 50 per cent of the value is in free gold, and this proportion is likely to remain stable as depth is attained. The sulphurets are reported to be of high grade. The old workings of Gold Hill extend up on the steep hill northeast of Quartzburg.

The Pioneer claim adjoins the Gold Hill on the south. A vertical shaft 400 feet deep is sunk on it. It is located on a fissured zone in the same porphyry dike which forms the hanging wall of Gold Hill, and which contains a great number of small seams with very rich ore, the whole forming a large body of low-grade ore. The porphyry is yellowish and decomposed, filled with pyrite, sericite, and some calcite. The sulphurets of the Pioneer are of much lower grade than those of Gold Hill.

A short distance below the Pioneer there are several claims on a vein crossing the creek, the principal one being called Mountain Girl. The ore consists chiefly of sulphides without any free gold, but sometimes contains much silver, being similar to the Centennial, which lies in a similar position in front of the Mountain Chief vein.

Immediately adjoining the Gold Hill on the northwest is a flat vein called the Lone Star, from which, in early days, much gold was obtained by surface sluicing.

The Iowa vein.—This vein lies a short distance north of the Gold Hill, and was opened in 1896 by an 800-foot tunnel starting from the 10-stamp mill one-eighth of a mile north of Quartzburg. The tunnel is driven through granite, which near the vein is much decomposed. The vein is a narrow seam in granite filled with a soft clay gouge and containing streaks extremely rich in gold. There is also some pyrite.

In the extension of the Iowa lie the Yellow Jacket and other claims which have been less developed.

The Carroll veins.—A group of five claims lies at the head of California Gulch, 1⅛ miles northeast of Quartzburg. The surface in this vicinity was extremely rich and has been extensively washed. It can not be said that the direction and extent of the veins have been definitely established, but that rich veins exist in this ground is not to be doubted. On the Ivanhoe claim some work was done in 1896, the vein being opened up by means of a tunnel several hundred feet long and exposing a good ore shoot. In this tunnel the vein dips south at a steep angle, and lies entirely in soft, decomposed granite. Outside of the ore shoot the vein is indicated only by a black-clay seam. The character of the ore in the shoot is illustrated in fig. 64, showing the breast of the tunnel. The valuable part consists of 2 feet of decomposed granite with seams 3 to 5 inches wide of pyrite. This contains considerable free gold, some of which appears to be in the decomposed wall rock. Outside of the principal shoot there are many streaks of iron pyrite, which do not contain any free gold.

The Kennebec claim.—This property, situated one-half mile from the Carroll, yielded some very rich ground for sluicing. Of the vein but little is definitely known.

Veins at head of Wolf Creek.—From the Iowa a string of claims extends through the Carroll and Kennebec and then farther through the Black Bear, Mountain Queen, and others up toward a high, prominent point on the divide. All these claims lie in granite, though occasional dikes cut the principal rocks; the main belt of porphyry lies a little to the south. On none of them has much work been

done. The Black Bear lies at an elevation of 5,000 feet a short distance to the west of the pass between Placerville and Garden Valley. The strike is N. 70° E., and the dip 50° to the south. The vein is a well-defined fissure with a pay streak of good ore 4 to 5 feet wide, which has been followed down for 40 feet. The ore is free-milling, at present at least, and contains a little galena. Northeast of the pass lie a number of more or less prominent veins. The Gold Dollar is a perpendicular vein, the ore of which carries no free gold, and it is said to run from $2 to $20 per ton. Near by lies another vein, with flatter dip of from 20° to 40°, carrying some free gold. The Monumental

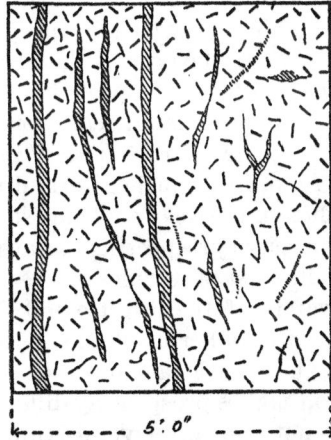

FIG. 64.—Breast of drift, Carroll veins. To the left, 2 feet of altered granite with rich seams of massive pyrite; to the right, 3 feet of altered granite with poorer seams of quartz and pyrite.

and Mountain Queen are on a nearly vertical, heavy vein striking N. 81° E. It has no distinct walls, but consists of streaks of quartz and heavy iron pyrite in decomposed granite cut by some porphyry dikes. The Etna, lying a short distance southwest of the pass, is said to consist of a streak in porphyry impregnated with auriferous seams and pyrite.

Veins in the porphyry dike east of Wolf Creek.—On the summit 1 mile east of the pass is the Golden Chariot, at an elevation of 5,300 feet. The wide dike of quartz-hornblende-porphyrite forms the country rock. In the tunnel the vein appears as a vertical streak, 2 feet wide, of brown decomposed rock, said to assay well. A similar deposit, called the Buena Vista, lies a little to the south. Half a mile eastward are the Big Six and the Mineral Hill group of claims, in the same porphyritic dike. The Big Six appears as a brown, much decomposed vein, 1 to 2 feet wide, chiefly made up of limonite; it strikes N. 55° E. and dips steeply toward the northwest, this dip being an unusual one for the vicinity. The claim is developed by a small shaft, and some fair assays were obtained from it, one sample containing $10.33 in gold and $0.50 in silver. The Mineral Hill group adjoins on the northeast, being situated on the headwaters of Ophir Creek, at an elevation of about 5,000 feet, and consists of five claims. The surface is extremely decomposed to a brown loam, and luxuriant vegetation covers all outcrops. These claims have not yet been prospected enough to determine their character, but the pay appears to be contained in streaks in the porphyry, impregnated with pyrite and carrying free gold on the surface at least. Near the Mineral Hill claim a large extent of surface is said to contain gold. The ore is a soft, decomposed rock, principally composed of limonite, one selected sample of which assayed $154 in gold, $1.71 in silver. Lead carbonate also occurs in the ore. Where sulphurets are found they are said to be of low grade. It is probable that these deposits are similar to that of the Pioneer claim, near Quartzburg.

Claims near Grimes Pass.—For 3 miles beyond the Mineral Hill group, in a northeasterly direction, no mineral deposits are known, though the deep soil covering the region makes it probable that no thorough prospecting has ever been undertaken. About three-fourths mile west of Grimes Pass, on the summit of the ridge dividing the waters of the Payette and Boise rivers, at an elevation of 5,000 feet, lie a number of claims called the Morning Star group. There are eight claims laid out along two adjoining lines. The deep surface soil makes prospecting difficult, and the exact character of the deposit is not known. Over a large extent the surface gives good prospects, and many little shafts demonstrate the presence of a considerable body of low-grade free-milling ore. The ore has always the appearance of streaks, 4 to 6 feet wide, of decomposed and brownish porphyry, striking a trifle north of east. The deposits lie in a dike, several hundred feet wide, of quartz-hornblende-porphyrite, continuing from the vicinity of Quartzburg, but the contacts of the dike with the granite are difficult to trace. Samples of the ore washed in pan gave good prospects of free gold, with some lead carbonate. On the Morning Star a shaft 232 feet deep has been sunk. A long tunnel has been started on the Payette side, 600 feet vertically below the shaft, and is calculated to strike the vein 1,700 feet from the mouth. In 1896 it had been driven as far as a point vertically below the shaft, but work had to be suspended on account of financial difficulties. Three hundred feet back from a point perpendicularly below the shaft the contact of granite and porphyry was struck in the tunnel. The deposits are evidently similar to those of the Pioneer mine and the Mineral Hill group, and consist of a more or less shattered zone in porphyry, which has been impregnated with auriferous sulphides by thermal action. The ore will probably be base in depth, but near the surface a considerable amount of free-milling ore exists.

The Mountain Queen mine lies on Grimes Creek, 3 miles above the Pioneer, at the southern edge of the porphyry dike and about one-fourth mile from the Morning Star claims. It is probably not the extension of the latter, but a more southerly vein. A few years ago a 20-stamp mill, driven by water power, was constructed on this property and ran two years. At first the ore is said to have been taken from a well-defined quartz vein, but later the mineralized porphyry was mined and milled and was found too poor for profitable work.

The porphyry dike extends eastward toward the high hills east of Grimes Pass, but was not followed and examined any farther. At Charlotte Gulch, on the east side of Grimes Pass, are many claims which on the decomposed surface carried much free gold. The veins contain pyrite, galena, and blende, with a small amount of free gold. Mr. Woods, of Placerville, states that there is evidence that those veins have been much disturbed by faulting.

MINING DISTRICTS EAST OF THE BASIN.

The Summit Flat mining district lies on the headwaters of Elk Creek, Clear Creek, and Moore Creek, at elevations of from 6,000 to 8,000 feet, and 12 miles north-northeast of Idaho City. At the headwaters of Elk Creek are the Barry, Peerless, King, and other veins, while the Wilson group of claims lie a little farther north. The veins generally strike east to west, or a little north of east, and dip south at steep angles. In character they are apparently well-defined, wide fissure veins carrying much quartz, chiefly free milling, though bunches of sulphurets may occur. The Mammoth claim in the Wilson group is opened by means of an incline 325 feet deep, exposing a considerable body of ore. There are two small mills in this mining district. This region was not visited by the writer in 1896.

Between Summit Flat and Kempner, 10 miles to the east, are many prospects with only slight developments, partly carrying gold ores, partly silver ores. Placer deposits occur at many places along Lost River and Bear River near Kempner. Twenty-two miles northeast of Idaho City the silver mines of Banner are located, which produced considerably before the recent fall in price of silver. Between 1882 and 1894 the total silver production of this district probably amounted to $1,500,000 or $2,000,000, reaching a maximum of over $200,000 in 1892. At present these mines are shut down. The deposits are large, well-defined quartz veins carrying rich silver sulphides.

Gold-quartz veins have been found on the southwestern slope of Sunset Mountain, and several claims have lately been located on

upper Rabbit Creek, draining into Boise River 8 miles east of
Idaho City.

Fineness of quartz gold.

Mine.	Value per ounce.	Fineness.
Gambrinus	$15.50	
Boulder	15.50–16.00	680 to 718
Forest King		700
Washington	15.00	
Illinois	15.00	
Ebenezer	16.50	
Gold Hill	a 17.50	800 to 910

a Average.

THE GEOLOGICAL HISTORY OF THE IDAHO BASIN.

The succession of geological events to which the existence of the
basin and of the gold-bearing gravels is due is neither simple nor easy
to decipher. In a large degree this is owing to the very monotonous
structure of the bed-rock series, which gives few clews, except those
indicated by the topography, to the character of the movements that
have taken place, for it soon becomes apparent, during a study of the
district, that erosion alone, unaided by orographic movements, can
not have produced this peculiar depression situated on the divide
between two main rivers.

The doubtful age of the granite, which alone constitutes nearly the
whole pre-Tertiary series in the basin, has already been alluded to.
It has further been stated that a surface laid through the ridge lines
of the Boise Mountains in general probably forms part of an old pre-
Tertiary peneplain or land mass planed down by erosion; and, still
further, that the erosion succeeding the uplift which differentiated
the Boise Mountains and the Snake River plains had, prior to the
lake period, cut far into this uplifted surface, so far, indeed, that
the Boise Canyon at its mouth was cut to its present depth at the
beginning of the Neocene. All this does not account for the depression of the basin, which lies much below the general level of that
surface. It seems probable that the present upper valleys of Grimes
and Moore creeks have been excavated by erosion, but this again
does not account for the basin as a whole. In it the river valleys
are separated by low ridges, the summits of which form, if extended,
an undulating surface considerably above the creeks, it is true, but
still much below the general surface of the surrounding country, as
is well shown by Pl. XCI. The extent of this surface determined
the existence of the basin in the first place, and as a probable working hypothesis to account for this it may be assumed that this earlier

basin was caused by a sinking of a portion of the Boise Mountains along curved fault lines between the Boise Ridge on the west, Wilson Peak and other elevations on the east, and the Thorn Creek Hills on the south. The age of the quartz veins in the basin can not be definitely indicated, but is most likely Cretaceous or Eocene. They were certainly formed before the deposition of the Payette lake beds, and it is to be expected that the rivers and creeks of the pre-Payette period of erosion in the basin carried detrital gold derived from these veins. These stream gravels are now either completely eroded or buried below the lake beds at Idaho City. As has been explained in the detailed description, there is slight chance of finding them by boring, and still slighter chance of mining them profitably if found.

The surface of the Payette lake attained a height above the present sea level of 4,200 feet at the mouth of the Boise Canyon. If the elevations and the topography were then the same as at the present time, the lake would have reached up as far as Centerville in Grimes Creek, and 4 miles above Idaho City in Moore Creek. It is almost certain, however, that the relative elevations are not the same, for near Idaho City we find the Payette lake beds at 4,400 feet, and there forming part of a smaller area which has settled down between parallel fault lines—just how much is not known. From this it appears that the basin has increased its elevation somewhat relatively to the country at the mouth of the Boise Canyon. High up near the divide, on Muddy Creek, a small remnant of inclined lake beds occurs, but it is perhaps a small local accumulation. At any rate it is certain that the Payette lake covered the lower part of the basin in early Neocene times.

The lake beds were rapidly accumulated in the bay then occupying the basin, and as no concentration of the material took place, it was natural that their content of native gold should be very slight.

The raising of the base level to the present elevation of over 4,000 feet would naturally produce extensive accumulations of gravel in the creeks draining to the lake. It is probable, indeed, that at this time of maximum lake extension the lower valleys opening into the basin were choked with gravel; and as an evidence of this may be cited the occurrence of auriferous river gravel on the summit of a ridge in the Thorn Creek drainage, at an elevation of 4,500 feet. Vast eruptions of basic lavas took place on the summit and western side of the Boise Ridge, and to less extent in the basin. All of these earlier lavas probably flowed out during the deposition of the lake beds. As the lake receded stream courses were established over its deposits, and the streams which headed near quartz veins began to carry down their precious load and concentrate the gold on the bed rock. These fluviatile deposits, which were formed very shortly after the Payette lake beds, or perhaps in part contemporaneously with them in valleys draining into the lake, have been described as "older gravels." Among them are the deposits of Gold Hill, East Hill, and Barker's claim at Idaho City, as well as the Ranch Company's claims and other gravels near Placerville. They rest partly on the lake beds, partly on granite; and while these gravels at Idaho City may have been deposited by Moore Creek as very high bench gravels, the gravels at Placerville form a part of a drainage system differing from the present one.

During the period of general erosion following the retreat of the Payette lake important events took place. The lower canyon of Moore Creek was scoured of its accumulated gravels. The lake beds were disturbed and acquired a decided dip to the west. One block of them at Idaho City sunk down between fault lines, being thus preserved from the erosion which destroyed the larger part of them. The gravels laid down on the lake beds also show a tilting westward, though at slighter angles than the latter. Near Placerville and Granite there were extensive disturbances, which greatly changed the old drainage. Along the eastern side of the Boise Ridge these disturbances, it would appear, took the form chiefly of faulting, for along this line the old gravels are cut off; but the amount of this faulting is not easy to establish. In this connection the occurrence of a little gravel on the summit of the Boise Range is of great interest, and it is difficult to explain. It is not believed, however, that the disturbance was so great as to create the whole of this ridge at this time.

Finally, at the close of the Pliocene came the eruption of the Snake River basalts. The Moore Creek flow originated at some point on Grimes Creek a few miles below the basin, and as it dammed the stream to an elevation of 100 feet the natural result was the accumulation of bench gravels above by the checking of erosion. As the creek gradually wore through the basalt filling, the level at which bench gravels were formed gradually sunk. Thus the bench gravels, lining the stream up to 100 feet above their beds, are directly due to the Pliocene basaltic eruption, and represent in the basin the deposit of the Pleistocene times.

CHAPTER V.

THE MINING DISTRICTS OF THE BOISE RIDGE.

NEAL MINING DISTRICT.

LOCATION.

The Neal mining district is situated south of the Boise River, just east of Three Point Mountain, on the head of Wood Creek, in Elmore County, 15 miles southeast of Boise, in the southwest corner of the Idaho Basin quadrangle.

The district embraces about 10 square miles, but the productive area has been confined to the heads of Wood and Bender creeks. The camp was discovered in 1889, and has been worked since during the summer months, producing about $200,000 in gold. Three mills have been erected, a 10-stamp for the Homestake, probably the oldest mill in the State, having been first used in Idaho Basin; a 5-stamp for the Alice mine, now a custom mill; and a 10-stamp for the Lilly mine on Black Creek, now idle.

Placer mining has been confined to a few bars and the creek beds, but the product from this source forms a very inconsiderable part of the camp's output. The largest amount extracted from the placers is said to have been $800.

The western half of the district is bare of timber, and except in the main streams there is during the summer months a scarcity of water. The eastern half is better timbered.

Pl. XCVII, drawn by Mr. F. D. Howe, shows the topography, dikes, veins, and mining claims of the central part of the district.

TOPOGRAPHY.

The main topographic feature is the high ridge which, with a WNW.–ESE. trend, divides the Snake River from the South Fork of the Boise, and which only reaches an elevation of about 5,000 feet. At Three Point Mountain this ridge swings to a northerly direction, and culminates 4 miles farther north in a point 5,400 feet high, between Birch and Grouse creeks, overlooking the mouth of Moore Creek and the forks of the Boise River. A number of deep canyons separated by narrow ridges radiate from the vicinity of Three Point Mountain. To the southwest extend, at the foot of Three Point Mountain, the Tertiary formations of the Snake River Valley. The direction of the creeks seem, in some measure, to follow the lines of faults and jointing—that is, they extend from north-northwest to south-southeast and from east to west.

GEOLOGY.

The prevailing rock is the normal gray granite of the Boise Mountains, composed of orthoclase, plagioclase, quartz, and biotite; hornblende is of rare occurrence in it. A jointed structure or sheeting is often noted, the direction (N. 50° to 80° E.) and dip (up to 45° S.) roughly corresponding to those of the veins. Numerous dikes cut the granite, and may be divided into several classes.

Some of the dikes consist of a harder gray granite, which carries some muscovite. Less frequent are dikes of coarser and more micaceous character than the general mass. Along the north side of Wood Creek are several dikes of a pyritiferous granite, occurring at one place as foot wall to a vein.

The most prominent dikes are those which crop so boldly on Black Creek, about 3 miles southwest of the Homestake mine, and which, according to Mr. Howe, continue for a long distance northward with a general trend of N. 5° W., one of them, the most easterly, showing on the map. Their width is up to 200 feet. The rock, which has a somewhat porous character, is dark gray in color with brownish spots; phenocrysts of feldspar are abundant, and are usually about 5^{mm} long. It is almost impossible to obtain fresh rock. The microscope shows the rock to consist of sanidine phenocrysts; small, brownish, decomposed foils and prisms, probably decomposed biotite; and a holocrystalline groundmass, of spherulitic and micropegmatitic character, of orthoclase and some quartz. The rock should probably be classed as a syenite-porphyry.

Normal granite-porphyry is common and forms dikes, more rarely irregular masses, with a general direction of N. 20° to 30° W., several occurring on the divide between Wood and Grouse creeks.

Lastly, there are narrow dikes of lamprophyres, dark-green, dense rock, in which small foils of black mica are often seen. These vary from 18 to 30 inches in width and often trend with the vein N. 78° E.

A specimen of this lamprophyric dike rock from the Hidden Treasure mine is a panidiomorphic granular rock composed of brown hornblende, augite, and soda-lime feldspar in slender, lath-like forms; probably also some orthoclase. The rock is very similar to certain camptonites, or, perhaps, stands between a minette and a camptonite. Similar dikes also occur in the Homestake mine, and in the foot wall of the High Five is a dike 15 feet thick, of lamprophyric rock with abundant black mica and porphyritic orthoclase crystals. These dikes are sometimes the hanging wall of the veins, at times apparently not affected by the vein processes, at others partially or entirely replaced by ore. At one place in the Homestake mine a part of this dike, crushed and altered, lies in the middle of a 4-foot vein. Mr. Howe concludes from his observations that the lamprophyres are the oldest dikes, followed by the large dikes of syenite-porphyry, and

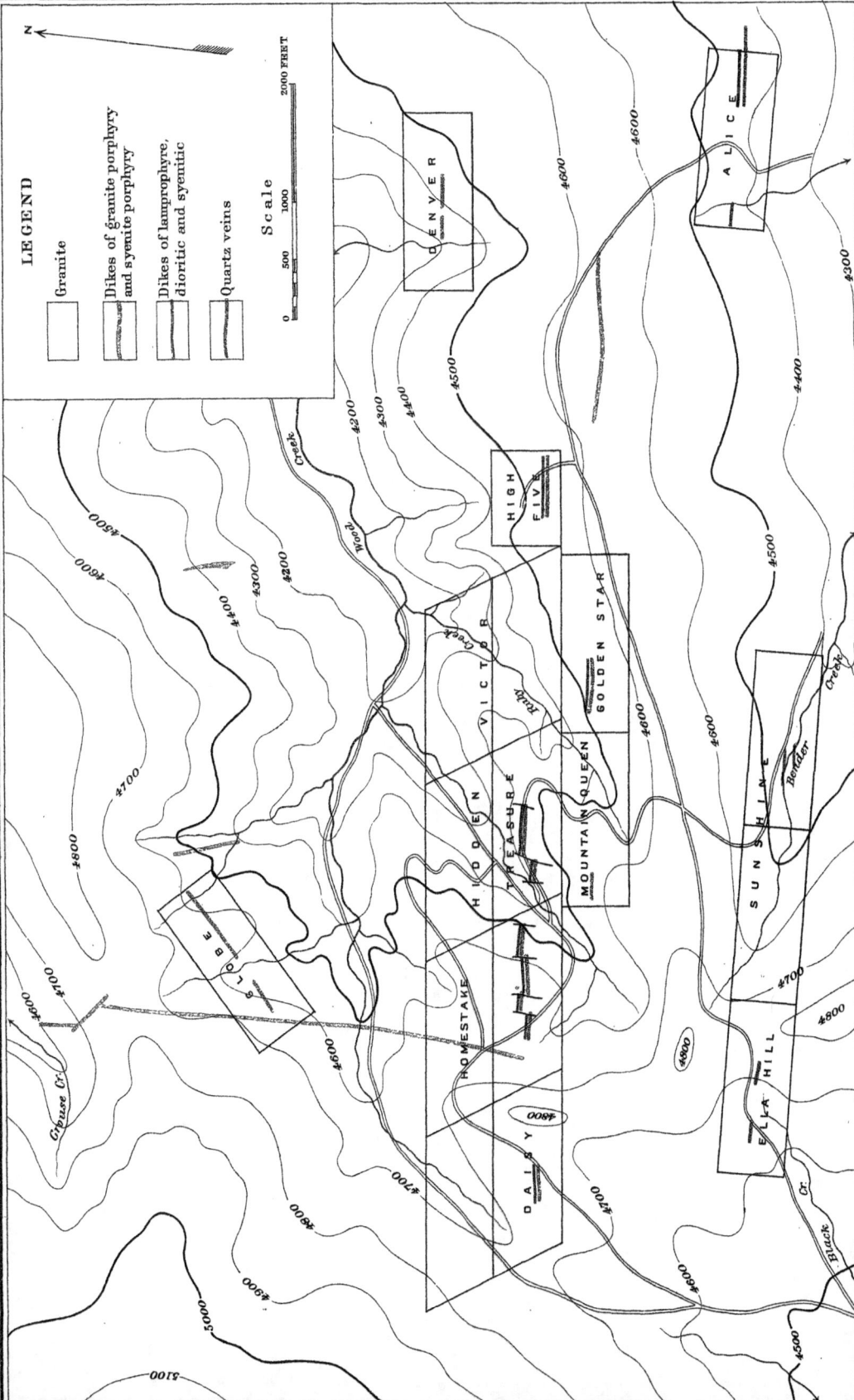

LEGEND

Granite

Dikes of granite porphyry and syenite porphyry

Dikes of lamprophyre, dioritic and syenitic

Quartz veins

Scale

0 500 1000 2000 FEET

these again by the granite-porphyry. It is probable that all of them
antedate the veins. Remains of a glassy rhyolite are found on Three
Point Mountain and the ridges northward.

THE VEINS.

The Neal district was visited in October, 1896, but only two days
could be devoted to it, and I therefore gladly availed myself of the
offer of the following notes by Mr. F. D. Howe, superintendent of
the Hidden Treasure mine, who is thoroughly familiar with the district and with whose statements my own observations in the principal
mines fully agree. Credit for many of the above data regarding the
dikes of the district is also due to Mr. Howe.

The veins of the district have the common N. 70° to 83° E. trend and the same
general dip to the south, and for form may be referred to three classes: First, veins
filling larger fault fissures, on which are located the principal mines; second, veins
along the minor shearing planes of the granite, more or less irregular, but generally carrying high-grade
ore; and third, veins of a
hard white quartz, called by
the miners "bull quartz,"
carrying no values.

The veins on the fault
fissures are often displaced
by faults of a north-south
trend, which occurred subsequent to the vein filling.
The position of the north-
south faults is often shown
on the surface by gulches of
greater or less size. The dip
is to the south, ranging from

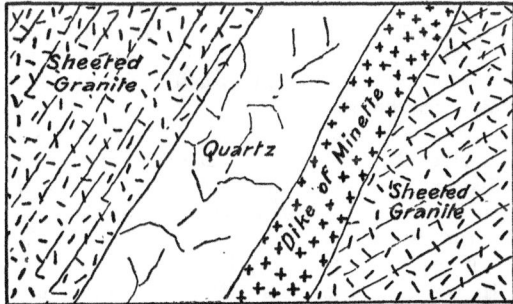

FIG. 65.—Cross section of vein in the Neal mining district.

30° to 54°, somewhat steeper than that of the foliation. The granite of the foot
wall for some little distance away from the vein has suffered a decomposition of
its mica, is harder, and is cut by cross jointing planes, in which are thin seams
of a talcose mineral, giving it a blocky appearance. Garnets occur on this side of
the fissure and may be of secondary origin. On the hanging-wall side the granite
is darker, and the main joints follow the fissure and no cross jointing appears.
On this side only appear the oxides of manganese. Beyond the sheeting or jointing of this side there is a zone of structureless granite (fig. 65).

The vein filling is separated from the walls by thin seams, but in places the
mineralization extends into both walls, in which case the gold in the walls is
coarser than that of the vein proper.

The vein matter consists of quartz, sulphides, and partly replaced country rock.
The ordinary structure is, next the hanging wall, a clear quartz, more or less
honeycombed and stained by the oxidation of the iron pyrites; next, a zone of
replaced country rock, granite or one of the dark rocks described above, carrying much pyrite and other sulphides.

Pyrite often occurs in the dike rock in detached kidneys and sometimes as par-
tial replacements. The width of the veins varies from 2 to 13 feet. The gold
occurs free in the quartz; partially free in the pyrites. It varies from microscopic

to shot size, and in some of the very rich pockets the oxidation of the pyrites has left a semicrystalline form. It is worth $15 per ounce from the retort, the alloy being silver. The associated sulphides are iron pyrites and galena, and a little zinc blende appearing on the surface as iron oxides and cerussite. From the pure galena silver values as high as 160 ounces per ton have been obtained, one assay showing 0.7 ounce of gold and 44 ounces of silver. The pyrites contain as high as 21 ounces of gold per ton. An assay of clean zincblende contained 1.4 ounces of gold per ton. The pyrites constitute from 3 to 10 per cent of the ore as broken; the galena less than 1 per cent. From 40 to 65 per cent of the gold is saved by amalgamation. The values of the milled ore range from $10 to $120 in gold per ton, the larger part being between $10 and $35. Concentrates carry about 2.5 to 4 ounces in gold and 5 to 6 ounces in silver per ton.

The veins of the second class have less development, and with two or three exceptions have produced but little ore. They occur nearly parallel to the fissures at some distance on either side, and differ in having less dip, little or no complete replacement of the country rock, and less regularity to the ore bodies. Generally the values are high. Where they have been productive the veins occur along the line of dike contacts, as at the High Five and Golden Star. The foot wall is here a granitic porphyry and apparently lies parallel with the jointing or sheeting.

The veins of the third class are the most prominent in the way of outcropping, the hard white quartz having suffered but little from surface decomposition. Occasionally the quartz is heavily stained by the iron oxides, and in places scattering pyrites are found, but the values are low. In places along the sides of this "bull quartz" have been found streaks of high-grade ore, but there has been no development to determine whether the hard quartz caps softer ore, as it does in other parts of the State.

Several prospects are found northwest of Three Point Mountain, in Charcoal Ravine. The main developments have been on the Jackson property—a narrow dike, 8 to 12 feet, of quartz-porphyry, with quartz veins on the joint and contact surfaces. The quartz carries a coarse gold, and there is some impregnation of the porphyry mass.

In the central belt the principal vein is the Homestake-Hidden Treasure. On the Homestake claim one shoot of ore, varying in length from 75 to 125 feet and in width from 4 to 12 feet, has been mined through tunnels to a depth of 350 feet. To the west the vein is cut by a fault, striking N. 19° W. and dipping 60° E. Beyond this fault the vein was recently found again, thrown 100 feet to the south. Four hundred feet to the east, in which distance another shoot is opened, a section of the vein is faulted 200 feet to the north. This faulted section is another ore shoot, and is about 250 long. Beyond this, to the east, another shoot has been mined to a depth of 100 feet. The term "ore shoot" refers only to ores yielding $10 per ton and over; if made to include ores from $5 per ton and up, the vein so far as opened is practically one shoot. The greatest depth attained below the surface is 350 feet.

On the Hidden Treasure, the easterly extension of the Homestake, the vein has been opened for 450 feet, to a depth of 165 feet, varying in width from 1 to 13 feet. Several small displacements by north-south faults, between 4 and 12 feet, are shown, and one with a displacement of 60 feet. The entire top of the vein has been moved to the south on a fault plane dipping 6° or 8° to the northeast, between 60 and 100 feet. In one place this plane was filled with ore. About 90 per cent of the production of the camp has been from these two properties.

In the High Five and the Golden Star a vein has been opened, with the granitic porphyry for a footwall. It apparently dips with the foliation. The high-grade ores occur in lenticular masses of greater or less extent. The value or extent of the rest of the vein material has not been determined.

The vein is in width from 2 to 6 feet, and in form and character of ores is but little different from the Homestake vein.

On the Corder property (Sunshine) an ore shoot has been developed on a fault fissure, with the ordinary dip and trend; and 120 feet to the north a parallel vein, dipping 30° to the north, has been opened. Each vein averages 4 feet, and the ores are similar in character and value to the type ores of the camp.

On Indian Creek, 5¼ miles southeast of Neal, a vein has been opened on the Stevens and Beck properties for 6,000 feet along its trend. It is a fault fissure 4 to 8 feet wide and carries ores similar in character and value to those of Neal.

On Black Creek, 4 miles southwest of Neal, several veins have been opened and some little production made. Work on them has been done only at odd times. The values are high and the limited developments indicate strong veins.

Two miles east, along Wood Creek, are two recent developments, the North Star and the Clements mines, both being of the common type of the district. In the first named arsenopyrite occurs.

Milling charges have been $5 per ton; transportation, from $1 up, and the mill saving not over 55 per cent—that is, only the free gold—thus requiring $12 to $15 ore to pay the outside expenses only. To the smelter the charges aggregate $28 to $35 per ton. It would be safe to say that nothing less than $25 rock would return more than wages to the owner unless he had a mill. The average claim owner had either to make the claim pay or limit the development to the annual assessment, carelessly done. The work in the camp is a result of this condition and fully illustrates it.

To these full notes of Mr. Howe should be added only the statement that the alteration of the country rock adjoining the veins is of the same character as that at other places within the region described. The black mica is bleached, being converted to carbonates and to white mica. The feldspars are converted to an opaque white matter, which is neither kaolin nor talc, but sericite in extremely fine-grained aggregates. The quartz grains remain unaltered in general, and pyrite and arsenopyrite are often introduced. The principal value of the veins is in the filling—that is, in the solid quartz accompanied by sulphides. But the altered granite next the filling here sometimes also carries good value.

BLACK HORNET MINING DISTRICT.

TOPOGRAPHY.

This district and its continuation northward (the Deer Creek mining district) is on the eastern side of the Boise Ridge, on the slope of the ridge culminating in Lucky Peak. It is 8 miles east-southeast of Boise, and is situated at elevations ranging from 4,500 to 5,500 feet. The topography is very accentuated; deep, sharply incised V-shaped canyons score the slope of the Lucky Peak, draining to Boise River and to Moore Creek. Scattered timber covers the hillsides at higher elevations. The topography is indicated in a somewhat generalized way on the Boise sheet.

GEOLOGY.

The geological structure of the Lucky Peak ridge is very simple. It consists nearly entirely of the normal granite of the Boise Mountains. A number of dikes vary the monotony, but consist almost exclusively of light-colored granite-porphyry, which have in general a northwesterly trend and a width occasionally attaining 100 feet, but usually much less. Placer deposits hardly found place to accumulate in the steep gulches and cut no figure in the production of the camp.

MINERAL DEPOSITS.

A number of gold-quartz veins are found in the Black Hornet district; and they are not confined to the district. Scattered small veins occur in the granite between this region and the Neal mining district, 9 miles to the southeast. West and southwest from Lucky Peak scattered prospects are also found; likewise to the north and northwest, connecting in the latter two directions with the Shaw Mountain and the Boise districts. The northern part, or the Deer Creek district, has been known for a long time. On the Montana claim an arrastre was built and worked many years ago by Mr. Plowman, of Idaho City. But the southern and lately most productive part, near the Black Hornet mine, has been known only during the last few years.

The production has been confined almost entirely to ore shipped to smelting works from the Black Hornet or Ironsides mine. During 1895 and 1896, 200 carloads are said to have been shipped, averaging $40 per ton, which would give a total product for the camp of about $24,000; the total production is probably $30,000.

The veins in this vicinity differ markedly in direction and dip from those at other camps on the range. Instead of a strike ranging from east-west to northeast-southwest, we here find veins striking north-south or northwest-southeast, and with a dip of 45° to 50° to the west. Though base ores prevail, some of the veins carry a notable percentage of free gold.

The more prominent claims begin at the Viola mine and extend for $2\frac{1}{2}$ miles northward. South of the Viola are a number of prospects, some of which, such as the Fraud and the Ruby, are reputed to be promising.

The Black Hornet vein extends through the Viola and Ironsides claims, but can hardly be traced any farther. Having at first a direction of N. 20° W., it changes in the Ironsides claim to N. 40° W., the dip being to the southwest at 50°. The vein crops along a ridge leading up to Lucky Peak, and a sharply cut ravine several hundred feet deep offered excellent place for tunnels to tap the vein. The Viola shows on the crumbling granite on the surface as a strong vein of white quartz. It is developed by a crosscut tunnel 200 feet long and a drift following the hanging wall north. A width of several feet of quartz is shown, and near the northern end of the claim a pay shoot exists 100 feet long and said to be 9 feet wide, carrying an ore which averages $15, of which about half is in free gold. The ore is is similar to that of the Ironsides. The Ironsides or Black Hornet adjoins on the north, and its principal pay shoot lies near the line of the Viola claim. It is developed by a cross cut striking the vein about 100 feet below the croppings and a winze sunk 100 feet deep ear the southern end-line in the drift. The pay shoot extends 200 feet north from the end line, but only 70 feet of it consists of shipping ore. Along the pay shoot the quartz reaches a width of 10 feet or more. This block of ore between tunnel level and surface was stoped and shipped, averaging, it is said, $40 per ton, almost entirely in gold.

North of the richest pay shoots are large amounts of lower-grade ore. The vein shows as a body of massive, fine-grained white quartz, from 2 to 10 feet wide, and contains sulphides irregularly distributed through it. The sulphides, which in the pay shoot will amount to 8 per cent of the ore, consist of arsenopyrite, pyrite, and zinc blende. The value is by no means exclusively in the sulphides, for a specimen of massive zinc blende and arsenopyrite assayed only 0.40 ounce of gold and 4.60 ounces of silver; a total value of $11.50. A sample of the quartz with scattered iron pyrite yielded 0.45 ounce of gold and 1 ounce of silver; a total value of $10. The walls are often ill defined and without a clay selvage, and consist of shattered granite and granite-porphyry altered by thermal processes. The feldspar is largely converted to sericite or white mica, and the rock contains, for a few feet on each side of the vein, much scattered arsenopyrite. This altered wall rock contains, in strong contrast to the filling, only a trace of gold and silver. The Ophir vein lies one-half mile north of Ironsides and has a similar direction. But little work has been done on it. Near the Ophir vein a long vein begins, and extends, with a westerly dip, due north across Dead Dog Creek to Deer Creek. The following claims are located on it, beginning at the southern end: McIntyre, Gray Eagle, Sorrel Horse, Golden Rule, and Montana. Most of them are but superficially developed, and the ore, though free milling on the surface, grows base at a slight depth. The Montana is opened by a cross cut 270 feet long.

BOISE MINING DISTRICT.

At a distance of from 3 to 5 miles east and east-northeast from Boise are a number of prospects which have never produced much, yet are worthy of mention. The country rock is normal granite throughout, cut by a few dikes of granite-porphyry with a general north-northwesterly direction. Dikes of dark lamprophyric rocks also occur.

Three miles east of Boise, on the north side of the stage road in Cottonwood Creek, are several claims, Morning Star, Last Chance, and First Chance, located on a narrow vein striking northeast-southwest and dipping south, the developments consisting only of two tunnels 100 feet long. High assay values in silver have been found on the first claim, while the others chiefly contain gold. One and a half miles further east, just north of the road, is another claim said to have yielded some rich decomposed silver ore.

On Picketpin Gulch, 5 miles east of Boise, are the Golden Star location and a great number of other claims. The Golden Star claim is said to cover two parallel veins and a cross vein, but the openings do not show the character of the deposit very clearly in the decomposed granite.

An arrastre was built on this claim many years ago and the soft decomposed ore treated in it is said to have yielded $33 per ton. A small mill was erected a few years ago, but it did not run long. Many of the veins in this vicinity show no quartz, but only a streak of thermal alteration on each side of a fault plane. Mr. Eldridge [1] mentions at this same locality "several narrow dikes of lamprophyre, trending N. 50° to 70° W. and dipping 45° to 80° SW. A vein of quartz lies between two of these." The dark-green, fine-grained dike rock from this locality is panidiomorphic granular, and consists chiefly of brown hornblende and orthoclase with some soda-lime feldspar. It appears to be a syenitic lamprophyre connected with the vogesites. In the lower part of Fivemile Creek many strong quartz veins appear, all of them having an east-west direction and a southerly dip of from 50° to 60°. The Scorpion is a well-defined vein showing 2 to 3 feet of solid quartz between granite walls. A tunnel 100 feet long has been driven in this vein in the western side of the creek. The quartz, which contains scattered iron pyrite and arsenopyrite, is said to assay up to $8 per ton. A dike of fine-grained, dark-green minette, a species of syenitic lamprophyre, was cut in the lower tunnel and appears to lie nearly parallel to the vein. The dike is considerably altered, filled with calcite, sericite (white mica), and pyrite, and carries about $1.65 in gold. Claims adjoining on the same vein are the Elevator and Hattie, while parallel to it and adjoining northward are the Badger and Free Gold. Parallel veins also lie on both sides of the Idaho City stage road at the mouth of Fivemile Creek, and scattered prospects extend eastward to Shaw Mountain district. The Tornado and Blizzard claims are situated 1 mile northeast of the Scorpion, and carry heavy sulphide ore, zinc blende, galena, and pyrite, having high assay value but containing no free gold.

---

[1] Sixteenth Ann. Rept. U. S. Geol. Survey, Part II, 1895, p. 235.

SHAW MOUNTAIN MINING DISTRICT.

The veins in this district, apparently forming a continuation of those on Fivemile Creek, were discovered in 1877. The veins are located on the high ridge just north of the Idaho City stage road, 8 miles N. 80° E. from Boise, at elevations of about 5,000 feet. The mines were prospected only on a small scale in 1896. The country rock is normal granite, with a few smaller masses of granite-porphyry. A strong sheeting of the granite is noted in many places between Fivemile Creek and Shaw Mountain, the joints having the same strike and dip as the veins. The Rising Sun vein is the most prominent, and crops as a well-defined fissure vein of white quartz for 1 mile near the summit of the ridge with a general east-west direction and dip of 45° to 80° S. Four claims are located on it, from west to east, as follows: Rising Sun, Paymaster, North Star, and Daisy. On the first of these a 10-stamp mill was erected in 1879. It is developed by several tunnels, the lowest 400 feet below the croppings and 500 feet long. From the upper levels at least 500 tons were extracted, yielding from $14 to $100 per ton; in the lowest tunnel heavy sulphide ore was found which did not contain much free gold. From a sample of this, collected from the dump and containing pyrite, arsenopyrite, blende, and galena, an assay of 11 ounces of gold and 4 ounces of silver was obtained, a total value of $230 per ton. The pay shoot is stated to be 120 feet long, the width of the vein being not over 2 feet. Some ore of similar character has also been extracted from the Paymaster. The North Star, showing strong croppings of white quartz on the summit of the sharp ridge, is said to contain five smaller shoots of ore, and is opened by a tunnel 280 feet long, giving backs of 180 feet. A few hundred feet south of this vein, on two smaller parallel veins, lie several claims, among which are the Kessler, Gold King, and Levi. These veins, on which some good ore is said to have been found, carry the same clean white quartz as the North Star; samples of this barren-looking material gave $2 in gold and $0.49 in silver.

Near the vein the granite has undergone the usual alteration due to thermal waters. The brown mica (biotite) is converted to white mica (muscovite), and the feldspars are changed to a white opaque mass, which is muscovite (sericite) in an extremely fine-grained aggregate. Aggregates of coarse muscovite sometimes occur in the quartz from this vicinity.

MINING DISTRICTS OF WILLOW CREEK AND ROCK CREEK.

LOCATION AND TOPOGRAPHY.

The Willow Creek district lies in Boise County, 18 miles distant from Boise, in a direction N. 20° W., and is adjoined on the northeast by the Rock Creek district, the two extending in an east-northeast direction for 8 miles. The elevation ranges from 3,000 to 5,000 feet above sea level. A ridge with northwesterly direction and culminating in Crown Point (elevation 5,300 feet) separates the two districts and also the watershed of the Boise from that of the Payette. Most of the mines of Willow Creek are located in the steep gulches at the head of the North Fork of the same creek. On the eastern side a steep descent leads down to Rock Creek, draining due northward into the Payette. The eastern end of Rock Creek district lies on the northeasterly trending ridge separating the Payette from the branches of Shafer Creek. The mining towns of Pearl and De Levan are located on Willow Creek. During the past summer there were probably 150 men in the districts. A small Huntington mill is erected on the Easter claim, and two smaller custom mills are also built lower down on Willow Creek.

While placer deposits were worked in the vicinity of Willow Creek long ago, and one of the mines of the district, the Red Warrior, was worked in 1870, the majority of the locations have been made during the past few years, especially in 1894 and 1895. The productive mines of the districts are at present those in Willow Creek, and the total output is probably about $80,000 in gold, all extracted within the last few years. The output of 1896 is estimated at $50,000. Among the producers are the De Levan group, the Checkmate, Friday, Leviathan, Birthday, and Lincoln.

GEOLOGY.

GRANITIC ROCKS.

The formation of chief importance as containing all the veins is the granite and associated dike rocks. A few miles west of Pearl the first granite hills emerge from under the cover of the Payette sandstones and rise rapidly eastward, extending thence uninterruptedly in a northeasterly direction. The larger part of the area is occupied by a granitic rock, which is a local modification of the normal granite of the Boise Mountains.

The rock crumbles and disintegrates very easily, and the slopes of the mountains are covered by a deep mantle of disintegrated rock. During the winter rains the erosion goes on very rapidly, and deep gulches are often excavated in a remarkably short time. It is possible to obtain the fresh rock only from exposures in the bottom of the canyons or in prospect tunnels. The rock is of a light-gray color and has a coarse-grained structure, the average grain having a diameter of from 3 to 6^{mm}. It is composed of white feldspar, quartz, biotite, and sometimes hornblende. Brown titanite in small grains is universally present. Occasionally larger grains of light-red orthoclase appear among the prevailing white plagioclase. Under the microscope the following minerals are noted: Hornblende appears in brown-

LEGEND

PLEISTOCENE

Pa

Alluvium and
River terraces

NEOCENE

Np

Payette formation
Sandstone etc

Nb

Basalt

Nr

Rhyolite

gr

Granite

PRE-NEOCENE

dr

Diorite Anglochurite

pr

Diorite Porphyrite

Gold Veins

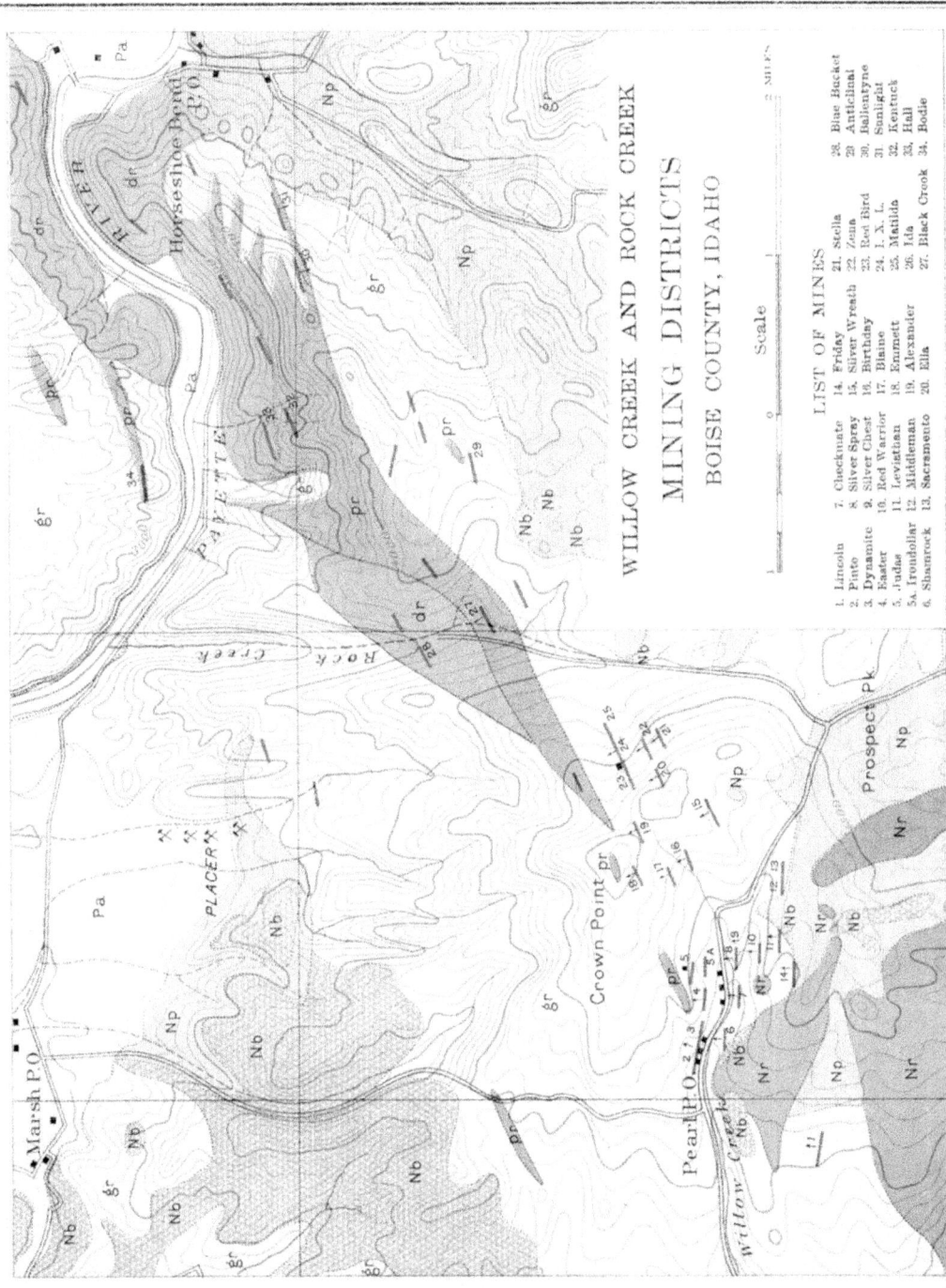

WILLOW CREEK AND ROCK CREEK
MINING DISTRICTS
BOISE COUNTY, IDAHO

Scale

LIST OF MINES

1. Lincoln
2. Pinto
3. Dynamite
4. Easter
5. Judas
5A. Irondollar
6. Shamrock
7. Checkmate
8. Silver Spray
9. Silver Chest
10. Red Warrior
11. Leviathan
12. Middleman
13. Sacramento
14. Friday
15. Silver Wreath
16. Birthday
17. Blaine
18. Emmett
19. Alexander
20. Ella
21. Stella
22. Zoza
23. Red Bird
24. I. X. L.
25. Matilda
26. Ida
27. Black Creek
28. Blue Bucket
29. Anticlinal
30. Ballentyne
31. Sunlight
32. Kentuck
33. Hall
34. Bodie

ish-green anhedral grains, and biotite as irregular, yellowish-brown foils. The quartz occurs in abundant anhedral grains, frequently exhibiting undulous extinction, due to pressure. The feldspar, also in irregular grains, is quite predominantly a soda-lime feldspar, generally an oligoclase or andesine, though some labradorite was found in a specimen from the Payette River Canyon. Small plagioclase crystals are sometimes embedded in the biotite. Orthoclase and microcline both occur in the specimens from the Silver Wreath mine and near the Checkmate, but are practically absent in other specimens from the Payette River Canyon, 3½ miles east of Marsh. A little magnetite and apatite always occurs. Titanite is present in larger quantities, and sometimes, as at the Silver Wreath mine, makes up a notable percentage of the rock. In this rock it occurs in idiomorphic wedge-shaped crystals protruding in feldspar grains and also including small prisms of the same mineral. An analysis of the rock from the Silver Wreath mine is given below, and a calculation of the analysis may be found on page 641:

Analysis of rock from the Silver Wreath mine.

[Analyst, George Steiger.]

SiO_2	65.23
TiO_2	.66
Al_2O_3	16.94
Fe_2O_3	1.60
FeO	1.91
MnO	trace
CaO	3.85
BaO	.19
MgO	1.31
K_2O	3.02
Na_2O	3.57
H_2O below $100°$	.18
H_2O above $100°$	.88
P_2O_5	.19
CO_2	.25
	99.78

According to these data, it is clear that the granitic rock is more closely related to a quartz-mica-diorite than to a granite, and shows great similarity to the granodiorite of the Sierra Nevada in California. Being, however, only a local modification of a large area of normal granite, with which it is connected by transitions, the name granite has provisionally been retained. More basic dioritic rocks, containing an abundance of hornblende, sometimes appear as irregular streaks and masses near Pearl. This is well shown in the tunnel at the Easter mine, as well as in the bottom of the creek near the Silver Spray claim.

A still more basic facies occurs at Horseshoe Bend on both sides of the river, and extends for a distance of 3 miles down along the canyon, bordering on the south against the porphyritic dike subsequently described and on the north against the dioritic granite, with very indistinct and ill-defined outline. This rock varies much in appearance, from dark-gray, medium-grained, or slightly porphyritic to coarsegranular, the latter consisting apparently of white feldspar and rather abundant green hornblende. The peculiarity of these dark, granular rocks is that besides being poor in quartz they also carry augite, and may be designated pyroxene-diorites. The rock at the Horseshoe Bend bridge carries imperfectly idiomorphic augite and a little hypersthene, the crystals being of a maximum length of 1^{mm}, but usually less. There is also a little biotite. These three constituents are embedded in a clear feldspar mass, which consists of short prisms of labradorite, with zonal extinction, cemented by smaller anhedral grains of probably the same feldspar. The structure in this, as in other specimens, is hypidiomorphic granular. The coarse-granular rock three-fourths mile west of Horseshoe Bend consists of white feldspar, green hornblende, and a smaller amount of biotite in large sheets. The microscope shows a few large, irregular grains of orthoclase, in which are embedded smaller prisms of labradorite. In this specimen there is a little quartz between the large feldspars. Most of the feldspar grains doubtless consist of labradorite. The augite is largely converted into hornblende, and but little of the original mineral remains.

DIKE ROCKS.

A series of dikes extends diagonally across the districts. Beginning as narrow dikes near the Horseshoe Bend post-office, these rocks extend in a widening belt over toward Rock Creek, there attaining their maximum development at the crossing of Rock Creek. The belt is here fully one-fourth mile wide, and practically forms one dike, though with smaller included masses of diorite. From Rock Creek this same dike extends up toward Crown Point Hill, but gradually contracts and ends before reaching the summit. Scattering dikes are, however, found on that hill, and appear to continue from there in a southwesterly direction, the last prominent dike appearing near the Dynamite claim at Pearl.

The dike rocks are of somewhat varying character. Pegmatitic dikes hardly occur at all, and light-colored granite-porphyry, so common elsewhere, is not abundant here. The usual dike rock is a coarse diorite-porphyrite. The rock occurring on the river hill between Horseshoe Bend and Rock Creek is prominently porphyritic by large white feldspar crystals, up to 1^{cm} long, and by smaller crystals of hornblende and biotite, all embedded in a reddish groundmass. Under the microscope the feldspars are shown to be a plagioclase of medium basicity. The hornblende and the biotite are of normal character, while the groundmass is microcrystalline, consisting of quartz and unstriated feldspar. The structure of the groundmass often is approximately micropoikilitic. In certain of these porphyrites hornblende is very abundant and occurs also as small prisms embedded in the groundmass.

Another kind of porphyry contains quartz as rounded phenocrysts, and in this variety no quartz occurs in the groundmass. At the crossing of Rock Creek the dike is wide and fresh and the rock is of somewhat different character. It is dark green and fine grained, with feldspar prisms up to 8^{mm} in length and a few black, shining crystals of hornblende. The feldspar is chiefly labradorite in sharp, short prisms and very fresh. The crystals are of all sizes, grading down to those which form a part of the groundmass. Augite occurs as idiomorphic crystals, many of them decomposed; a little hornblende is also present, and magnetite is quite abundant. Between the closely crowded feldspars lies very little groundmass of quartz and unstriated feldspar. The structure of this rock is intermediate between holocrystalline porphyritic and panidiomorphic granular. The rock is thus an augite-diorite-porphyrite, closely connected with the lamprophyric dike rocks.

At the Dynamite claim, near Pearl, dense, dark-green dikes appear, consisting of clouded and altered feldspar in lath-like form, between which lie the decomposed ferromagnesian silicates replaced by chlorite and epidote. This rock has somewhat the appearance of an altered, fine-grained diabase. Scattered dikes occur in other parts of the district, one large dike cropping on the road to Marsh, 1$\frac{1}{2}$ miles north of Pearl. On the whole, the dikes may be said to be most abundant for a distance of half a mile north and half a mile south of the main belt of porphyritic rocks above described.

THE PAYETTE FORMATION.

The loose sandstones of the Payette formation (early Neocene) are laid down upon the very irregular granitic surface and begin at the western edge of the Willow Creek district, extending far westward. Sandstones and fine gravels of the same age are also found high up in the vicinity of Prospect Peak, and cover, in fact, the pass leading from Willow Creek to Boise. A branch of the Payette sandstones extends in a northeasterly direction as far as Horseshoe Bend. The Payette formation locally contains gold placer deposits, as at Church's ranch, at the southern edge of Marsh Valley. The gold in these placers has doubtless been washed down from quartz veins on the northern slope of Crown Point Hill. Other placer deposits occur 1 mile southwest of Marsh in the sands and conglomerates of the Payette formation.

These placers also probably have their origin in the gold veins of Crown Point Hill.

RHYOLITE.

The sharp point of Prospect Peak and the hill 1 mile to the west of it consist of rhyolite. They are necks from which large flows of the same rock poured down the southern slope of the ridge during the period of the Payette lake beds. In South Willow Creek the sandstone is seen to overlie the rhyolite. A smaller flow of rhyolite reached down as far as one-fourth mile south of Pearl. The rock is usually reddish or reddish-gray, and is of the ordinary compact lithoidal variety. In the last-mentioned flow occur, associated with it, rhyolite glass and loose tuffs.

BASALT.

The eruption of rhyolite was followed, during the same early Neocene period, by extensive eruptions of andesitic basalt. Smaller masses and necks of this black massive rock occur on the ridge one-fourth mile west of Prospect Peak, near the Leviathan, and at several other places to the west.

The Payette formation and accompanying eruptives are later than the mineral deposits, and contain no veins.

THE ORE DEPOSITS.

GENERAL CHARACTER.

The gold deposits in the Willow Creek and Rock Creek districts are fissure veins of somewhat varying character. Most of them occur in a belt parallel to that of the porphyry dikes, extending in a northeasterly direction, and being in no place much over 1 mile in width. The veins at Willow Creek are most frequently entirely in the dioritic granite. Sometimes a vein follows a porphyry dike for some distance in foot or hanging, but rarely for a long distance. Again, a vein may cut through a dike, in which case it often splinters up. The porphyry dikes are evidently all older than the veins. The fissures which carry gold strike east-west or northeast-southwest. In the Willow Creek district the dip is always to the north from 45° to 80° and the direction east-west, but toward Rock Creek the direction gradually changes to northeast-southwest. Toward Horseshoe Bend the direction changes again to east-northeast to west-southwest, and the dip is frequently steep to the south. The individual veins can rarely be traced for a long distance, and though it is probable that some of the veins are a mile long, this can rarely be satisfactorily proved. Narrow veins predominate in the Willow Creek district, while wider deposits occur on Rock Creek. The best exposures are, however, found in the former district, owing to more extensive development. In Rock Creek the developments are relatively slight, and the decomposed surface material does not always allow satisfactory conclusions as to the character of a deposit. A large number of locations have been made in both districts, which, in fact, contain a very great number of veins. The ordinary type of the Willow Creek deposits consists of one or more fault fissures, on both sides of which there is a zone a few feet wide in which the country rock has been thoroughly altered (see Chapter II) and impregnated with pyrite. Along the main fissure, or, if there are two or more, chiefly along the foot wall, there are narrow seams filled with sulphides (pyrite, zincblende, arsenopyrite, and galena), which constitute the ore. The altered country rock, though often studded with pyrite crystals, usually contains only $1 or $2 in gold, while the value of the solid sulphides in the seams may reach $100 per ton or more. The deposit may thus be characterized as narrow veins of high-grade sulphide ore. There is usually but little gangue along these seams; calcite and quartz both occur. In other deposits the zone of altered diorite or granite is traversed irregularly by numerous small seams carrying arsenopyrite, blende, and galena, and in this manner the wider deposits of medium-grade ore are formed. The veins of Rock Creek are generally wider than those of Willow Creek.

The surface decomposition attains 50 to 100 feet in depth. In this zone the vein matter forms a brown ferruginous mass, which contains free gold and partly decomposed sulphides, often also lead carbonate. In many cases the fresh sulphide ore is found less than 50 feet from the surface. The fresh ore contains a very small percentage of free gold, and sometimes, in rich ore, not even a color is obtained by the pan. It follows from this that the amalgamation process is, as a rule, applicable only to the surface ores, though some veins will be found to contain more free gold than others. At present the rich ore—all above $30—is sacked and shipped to smelters.

The minerals consist of the following combination, so common in the Boise Ridge: pyrite, arsenopyrite, zinc blende, and galena; chalcopyrite is rare. The first two often occur as crystals. The zinc blende is black, brown, or greenish-yellow, usually not crystallized. The galena is less common than the others, and is considered an indication of rich ore. Ruby silver is reported as a rarity from the Shamrock and the Lincoln.

Shipping ore often contains 5 ounces of gold and 5 ounces of silver to the ton. A sample of pyrite, arsenopyrite, and galena from one of the best mines gave 0.85 ounce of gold and 28.35 ounces of silver per ton, a total of $37.42. Some galena carries 60 to 70 ounces of silver, and generally also much gold. Much of the arsenopyrite and zincblende is poor. The principal value appears to be in the pyrite and galena. Of the extent and direction of the ore shoots but little is known at present, but it is clear that the high-grade shoots are not of great lateral extent and that they are rather irregular.

TREATMENT OF THE ORES.

At present only the shipping ores can be utilized, and the question how to make $10 to $20 ore pay is one of the greatest importance to the camp. If it should contain any notable amount of free gold, amalgamation and concentration will probably be found most economical. Experiments should be made as to the applicability of the cyanide process to these ores. In the absence of notable amounts of copper and antimony the process might be of advantage, but experiments are necessary to prove this.

DETAILED DESCRIPTION.

The Lincoln vein is one of the most westerly locations in Willow Creek, being situated about a mile south-southwest of Pearl. The strike is N. 76° W. and the dip steep to the north, the vein being traceable in granite for a distance of one-fourth mile, with good ore at close intervals. The tunnel shows 2 to 10 feet of altered and pyritic granite, with smaller streaks and veins (1 to 6 inches wide) of, pyrite, arsenopyrite, and blende, associated with a little drusy quartz; ruby silver has been noted. The developments consist of a crosscut tunnel 200 feet long, with drifts on the vein, and a small winze. Fifteen tons of ore were shipped, averaging $100. A sample of poor ore gave 0.1 ounce of gold and 5.30 ounces of silver, a relatively large amount of silver for this camp.

The Shamrock is situated a few hundred feet southwest of Pearl post-office, and is developed by an 80-foot crosscut from the creek level, from which drifts extend on the vein. The country rock is very much disturbed granite containing rich but irregular seams of ore. Some tons of the latter were shipped to smelters.

The Pinto lies a short distance north of the Bishopric mill, in Pearl, and has been opened by 225 feet of tunnels and a 50-foot winze. Sixty tons of ore are said to have been milled, yielding $23 per ton. The vein is about 2 feet wide. The Pearl claim, showing some very good ore, is located in this vicinity.

The Dynamite, said to be the extension of Pinto, is situated on the north side of Willow Creek, a short distance east of Pearl, the croppings being 250 feet above the stream. The vein is opened by a tunnel 200 feet long, strikes N. 68° W. and dips 45° N., and shows 16 inches of decomposed vein-matter in granite. A large porphyry dike lies close by to the north.

The Easter lies very nearly in the continuation of the Dynamite, and is one of the producing veins of the district. The croppings lie in granite 250 feet above the creek and the vein is developed by a tunnel 100 feet below the croppings. Another tunnel was started at the creek level, but has not yet reached the vein, work being suspended in 1896 on account of litigation. A considerable amount of ore was stoped and milled in 1895. The vein strikes N. 81° W., dips 60° N., and shows 2 to 3 feet of decomposed vein matter which, in the pay shoot, yielded, it is said, $38 per ton of free gold. Pockets carrying extremely rich ore occurred at intervals. On the west side the vein is said to splinter up in a porphyry dike.

The Iron Dollar is located a short distance east of the Easter, and is probably the continuation of the same vein. The development consists of short tunnels and surface cuts. A few tons have been milled, yielding $58 in free gold. The vein, which strikes N. 71° W. and dips 60° N., lies in granite with occasional porphyry dikes in the hanging wall. The decomposed ore consists of altered granite, with streaks of arsenopyrite, pyrite, and galena, as well as a little calcite and quartz. A sample of good ore gave 2.50 ounces of gold and 7.15 ounces of silver to the ton, a total of $56.68.

The Judas lies a few hundred feet north of the Iron Dollar, and is at present (1897) being developed by a shaft intended to reach a depth of 400 feet. Excellent ore is reported to have been milled from this vein. The strike and dip are nearly the same as in the Iron Dollar, and the vein shows from 1 to 2 feet of decomposed granite carrying rich seams.

The Checkmate crops in granite on the south side of Willow Creek, due south of the Judas. This vein is one of the productive properties, shipping ore during the whole summer of 1896. It is reported that 300 tons, averaging $80 per ton, have been shipped. The mine is developed by a tunnel 100 feet long on the level of the creek, giving about 100 feet of backs. The deposit consists of a zone several feet wide of altered and pyritic granite, containing rich seams of heavy sulphides, arsenopyrite, pyrite, blende, and galena. At the time the mine was visited most of the ore came from a seam of solid sulphides 4 to 6 inches wide. The strike is N. 84° W.; the dip to the north.

A short distance east of the Checkmate is the Silver Spray, from which some good ore has been shipped. The vein is opened by a short tunnel, and strikes east-west, dipping 40° N. The country rock is dioritic granite, with streaks of dark diorite and dikes of diorite-porphyrite. The character and the minerals are similar to the Checkmate.

A little farther east is the Golden Chest, showing a wide zone of altered and pyritous granite with narrow seams of zinc blende.

South of Willow Creek, opposite the last-named claim, lies the Red Warrior, the oldest location in the district. No work was done on it in 1896.

One mile southeast of Pearl, on the summit of a ridge, lies the Leviathan claim. This is developed by a shaft 75 feet deep, and a few tons of rich ore have been shipped. The width of the mineralized granite is 13 feet, with seams of pay ore on both walls. In the continuation of this claim lie the Middleman and the Sacramento, which have the same east-west strike. Both of these claims have shipped some rich ore, containing much galena. The pure galena, assays 8 ounces of gold and 67 ounces of silver to the ton. The developments are slight.

The Friday is located on a flat 100 feet south of the Leviathan and is developed by a 100-foot shaft and drifts on the vein. Thirty tons of ore are said to have been shipped and some rich surface ore has been milled. The deposit shows 8 feet of altered granite, with a seam of heavy sulphides on the foot and hanging walls. The ore contains pyrite, arsenopyrite, and blende, with much calcite. This vein is said to be nearly vertical. The mine was not accessible during the present examination.

Beyond the claims mentioned, for a distance of about 1,000 feet, there are but few locations, but a number of strong veins are found on the summit of the ridge separating Willow Creek from Rock Creek.

Beginning on the north, the Emmett vein is located near the summit of Crown Point Hill, on the southwesterly slope. Good ore is reported to have been found in this claim and shipments were made during the winter of 1896–97. An incline shaft is sunk to a depth of 100 feet.

The Ida lies on the Rock Creek slope, at an elevation of 4,700 feet, and appears to be a wide vein in a dike of mineralized porphyry. It is developed by 175 feet of tunnels and shaft, and excellent assays have been obtained from average samples. The Blaine, on the Willow Creek side, a short distance north of the road to Rock Creek, is a promising prospect from which some ore was shipped in 1896. The Alexander, located on the divide, showed some good ore, a brown decomposed mass containing lead carbonate and milling $40 per ton. The vein is 4 feet wide, striking N. 74° E. and dipping north.

The Birthday, a few hundred feet south of the Blaine, has produced some rich shipping ore, composed of solid pyrite, zinc blende, and galena.

The Silver Wreath lies on the Willow Creek side, three-fourths of a mile southeast of Crown Point Hill, and is opened by a crosscut 170 feet long, cutting the vein at a depth of 75 feet. The croppings show distinctly by quartz, colored greenish by arsenic. The strike is N. 79° E. and the dip steep to the north. The deposit consists of a zone of decomposed dioritic granite, 8 feet wide, containing seams of the usual minerals. Some ore shipped yielded $40 per ton.

The IXL is situated on the Rock Creek side, three-fourths of a mile east-southeast of Crown Point Hill. An incline shaft is being sunk on this vein, and has at the present writing attained a depth of 200 feet. The vein, which strikes on an average northeast and dips 50° N., is contained in granite with a dike of granite-porphyry in the hanging wall. Four feet of brown decomposed vein matter are shown in the shaft. About 25 per cent of the total value is in free gold, and, according to average samples, there is a considerable body of medium-grade ore. A tunnel is located on the eastern extension, 900 feet east of the incline. It is claimed that several ore bodies occur between the incline and the shaft. The vein is one of the longest in the district, being traceable for at least 2,000 feet.

The Zena, Stella, and Ella claims are located 600 feet south of the IXL, about in the continuation of the Birthday and Silver Wreath, on the Willow Creek side, and are opened by several short tunnels. The Zena shows a well-defined fissure with a dike of diorite-porphyrite 5 feet wide in the hanging. The ore consists of narrow streaks of blende, pyrite, and arsenopyrite in an altered and pyritous granite. The ore appears fresh near the surface, in contrast to the decomposed ledge matter of the IXL and the Alexander.

A number of veins are located along Rock Creek. The Black Crook lies 1½ miles northeast of Crown Point Hill, and is opened by a drift on the vein 140 feet long. The vein strikes N. 62° E. and has a maximum width of 8 feet. It has diorite-porphyrite in the hanging wall and granitic granite-diorite in the foot wall. The gangue is a grayish quartz and a pink calcite colored by manganese. The ore body was reported to be large but of low grade. Some assays show a relatively large amount of silver in the ore. The vein is reported to be traceable for a long distance eastward.

The Blue Bucket lies 1,500 feet farther down the creek, at an elevation of 3,500 feet, in diorite and dioritic granite. Some very rich ore is said to have been taken from it, but the claim was not worked during the examination of the district.

East of Rock Creek lie a large number of claims, on most of which but little work had been done in 1896. Among them is the Anticlinal, under the Liberty Cap Hill, and the Lambertine, Bobtail, and Mint claims, three-fourths of a mile east of Rock Creek. On the Bobtail claim a tunnel 200 feet in length has been run, cutting an 8-foot vein of good ore.

Many claims are also located on the ridge between Shafer Creek and the Payette River. There appear to be two principal lines of deposits. One begins one-half mile southwest of Horseshoe Bend post-office, where the granite emerges from the Payette lake beds and continues in a west-southwest direction up to the summit of the ridge. Among the claims located along this line are the Sunny Side and the Ballentyne, both on the same vein. The vein dips 70° to 80° S. and is about 2 feet wide. The vein matter is soft and decomposed, carrying free gold and some lead carbonate. The country rock is granite, but in the foot wall lies a dike of quartz-diorite-porphyrite a few feet wide. Another vein lies a short distance southward. A large number of prospects are found on the steep river hill toward the Payette, about one-fourth mile north of the Ballentyne and opposite McFarland's ranch on the river. The following claims are located on this vein system, from east to west: Mammoth, Apex, Atlanta, Claggett, Topeka, Kentuck. These are at an elevation of about 1,000 feet above the river. Hall's claim lies a little lower down, about 700 feet above the river. Some of these claims appear to be promising properties, but the developments are slight. The ores are decomposed, carrying on the surface a considerable amount of free gold. Some antimonite, carrying no gold, was also found in this vicinity.

The above are by no means all of the claims and prospects of the district, but only such as showed any notable development or production of ore. It is quite possible that some of these prospects may develop into paying mines.

SILVER DEPOSITS.

The Boise Mountains contain many notable silver deposits, chiefly well-defined quartz veins with finely distributed rich sulphides and antimonides. But it is not intended to take up the study of them in this paper. The principal locality where silver mining has been carried on is at Banner, 25 miles northeast of Idaho City.

Many scattered quartz veins with silver ores, either galena or rich silver sulphides, occur in the area here described, but none of them have produced much. A few deposits of this kind occur along the Idaho City road 3 or 4 miles from Boise, and another in north fork of Dry Creek a few miles southwest of Shafer Butte. Other silver prospects are located 2 miles south of Church's placers, in Marsh Valley, and at several places near Horseshoe Bend, notably on the western side of the Payette 2 miles north of the bridge. Many silver prospects occur 1 mile east of Halfway House in the Moore Creek Valley, and some of them are said to contain rich ore (Sunlight group).

The occurrence of occasional silver deposits in the Idaho Basin has already been mentioned in the detailed description in Chapter IV.

PLACERS OF THE BOISE RIDGE.

RECENT PLACERS.

The bars of the Boise and Payette rivers were worked in the early days, and on some of them work is still progressing. The large gravels of the lower reaches of both rivers contain a little gold, but scarcely enough for profitable working. A dredger built some years ago to work the gravels of the lower Payette near Marsh proved a failure. The placers of the Moore Creek drainage were discussed in Chapter III. Most of the creeks of the Boise Range have carried a little gold, but few of them have been rich.

Benches along Dry Creek and Willow Creek are worked at intervals, even now, when water is available. Shafer Creek, at least the branch heading near Cartwright ranch, carried a little gold. The richest placers probably occur at the northern base of Crown Point Hill at Church's in Marsh Valley; but the whole output of the recent placers of the Boise Ridge is, if we except the basin, of small importance.

NEOCENE PLACERS.

The shore and old gulch gravels resting on granite in the early Neo-cene Payette formation carry a little gold at many places—for instance, in several gulches about 2 miles eastward from Boise, north and south of the Idaho City stage road. Similar old placer deposits are found in the Payette formation at Church's, in Marsh Valley, and the old grav-els at Johnson's, 1 mile southwest of Marsh, contain some gold which has been concentrated in the gulches and washed by the hydraulic process.

Gold-bearing gravels of late Neocene (Pliocene) age are found below the remnants of the several basalt flows on both sides of Boise River, those in Moore Creek having already been mentioned. The top of the lowest flow, which is probably the oldest, lies at the "New York House," 10 miles southwest of Boise, at the level of the river, and is not visible farther west. Eastward it rises slowly, and near the mouth of Moore Creek is 40 feet above the river (in August). Below this flow, which is about 20 feet thick, lie 2 to 10 feet of coarse, heavy gravel, resting on granite. This gravel is in places rich in coarse gold, part of which probably comes from seams and small veins in the surrounding gran-ite. There are only a few exposures of this low flow below the mouth of Moore Creek, and it is reported that only one or two are known above. The flow came down the south fork of the Boise. At low water this gravel below the basalt has been mined with profit at sev-eral places, notably at the Holy Terror mine, 2 miles below the mouth of Moore Creek, and at Tarents, 2 miles farther down. There is only a limited amount of this gravel below the lower flow. The two other flows, 30 to 60 feet thick, are at elevations of 120 and 300 feet above the river. Underlying both of them, wherever they are preserved, hanging along the banks of the canyon, are heavy masses of late Neo-cene gravels, 20 feet or more in thickness. This gravel contains some gold throughout, and though much of it is fine, it may in places be found rich enough for the hydraulic process wherever water can be brought to it. In 1896 an attempt was made at the mouth of the canyon, 8 miles southeast of Boise, on the northeast bank of the river, to mine the heavy mass of gravel and sand here underlying the basalt flow, and should this attempt be successful there are probably many other similar deposits a little farther up the river which could be worked in the same manner.

www.ingramcontent.com/pod-product-compliance
Lightning Source LLC
LaVergne TN
LVHW021517080426
835509LV00018B/2544